THAT'S DISGUSTING

ALSO BY RACHEL HERZ

The Scent of Desire

THAT'S DISGUSTING

Unraveling the Mysteries of Repulsion

RACHEL HERZ

 W. W. NORTON & COMPANY   NEW YORK   LONDON

For information about permission to reproduce selections from this book,
write to Permissions, W. W. Norton & Company, Inc.,
500 Fifth Avenue, New York, NY 10110

For information about special discounts for bulk purchases, please
contact W. W. Norton Special Sales at specialsales@wwnorton.com or
800-233-4830

Manufacturing by Courier Westford
Book design by Ellen Cipriano
Production manager: Louise Mattarelliano

Library of Congress Cataloging-in-Publication Data

Herz, Rachel, 1963–
 That's disgusting : unraveling the mysteries of repulsion / Rachel Herz. —
1st ed.
     p. cm.
 Includes bibliographical references and index.
 ISBN 978-0-393-07647-9 (hbk.)
 1. Aversion. I. Title.
 BF575.A886H47 2012
 152.4—dc23

                    2011041675

W. W. Norton & Company, Inc.
500 Fifth Avenue, New York, N.Y. 10110
www.wwnorton.com

W. W. Norton & Company Ltd.
Castle House, 75/76 Wells Street, London W1T 3QT

1 2 3 4 5 6 7 8 9 0

FOR LARRY

CONTENTS

Preface

My fascination with disgust was ignited in the spring of 1995 in a semi-nar led by Paul Rozin, "the father of disgust in psychology,"[1] at the University of Pennsylvania. At the time, I was a new faculty member at the Monell Chemical Senses Center, an institute dedicated to the senses of smell and taste which was located around the corner from Paul's lab. Preoccupied with the immediacies of my own research career on odor and emotional memory, I didn't pursue any of the fascinating tidbits that I heard from Paul at that time, but they percolated quietly in the back of my mind. Fast-forward thirteen years—and a confession that my first inclination to write a book about disgust began as a joke.

In March 2008 I was invited to be the celebrity judge for the National Rotten Sneakers Contest in Montpelier, Vermont, an annual contest held since 1975 that's notable enough to be listed in the *Farmer's Almanac*. I was told that my job would be to sniff the sneakers of kids from around the country who had already won their regional odorifer-ous challenges. The finalists, aged six to sixteen, would now be com-peting for recognition as having the stinkiest footwear of them all. In addition to a cash prize, the winner has their sneakers enshrined forever in a glass cabinet dubbed "The Hall of Fumes." The contest included other judges whose jobs were to look at various components of the shoes' exterior—laces, grommets, soles—in order to rate how rotten they were. But only two of us had to get up-close and nose-personal. Before I left for Vermont, many of my friends cajoled me with questions: "How will you be able to stand it?" "How could you have agreed to such a thing?" "Won't it be just too disgusting?" I started to joke with them that I was doing it as research for the sequel to the book I had just written, *The Scent of Desire*, and that now I was going to write "The Scent of Disgust." This is not quite what happened.

The stench of the sneakers, though far from mild, wasn't nearly as bad as I thought it would be. I've even been the nose judge in the years since and have agreed to be on-sniff for as long as the contest organizers may want me. Why wasn't it so bad? The answer in part was because I had psyched myself up to believe that it would be excruciating, and the reality was weak by comparison. The fact that my thoughts enabled me to tolerate those mephitic sneakers made me realize that our mind has a very powerful influence on our perception and experience of disgust. And this got me thinking more. I started to read papers and books and go to scientific conferences all starring the emotion of disgust and what I discovered is that disgust is incredibly fascinating and complex—and not yet fully understood.

I am a research psychologist who has been studying emotion, perception–cognition, and especially the sense of smell since 1990. I have always had a particular fascination with disgust, partly because it shares many commonalities with the sense of smell. They are both biologically based, but enormously influenced by learning, context, and complex thought. However, I did not begin to study disgust in earnest until 2008. In the time since then, I have researched, thought about, experimented on, and written about disgust nearly continuously.

Over the past few years, I have been taken on an exploratory odyssey like no other. I have learned that the emotion of disgust is universal, but it is not innate. I have learned that disgust is many things to many people at different times and in different places; and that it is shaped by our situation, culture, and personal history. I have learned that disgust is uniquely complex among human emotions. And I have learned much more. Coming to the topic of disgust as a relative newcomer, I believe that I have the same basic questions and curiosity that you will have about it, and my scientific background and training have enabled me to decipher many of the mysteries of disgust and arrive at some novel conclusions.

This book not only explores what disgust is, but also how, where, and why it is elicited psychologically and neurologically, what its dark sides are, our perverse attraction to it, and its consequences for us as individuals and for our society. Drawing on scientific experiments, news headlines, experiences from my own life, and the stories of others, I will unearth a host of thought-provoking facts: for example, that there are specific rotted foods that every culture, including ours, considers a delicacy; that homosexuality is at once practiced and punished by death in the same country; and that women dislike foreigners more when they are pregnant. You also might be surprised to learn that some people can't experience disgust, that incest and cannibalism are upheld in some traditions as part of the privileges of royalty or the rites of religion, that publically relieving yourself in the dining room was once perfectly normal, and that necrophilia is not a crime in many American states. You will also discover how politicians can use disgust to win votes, how disgust can be managed in the courtroom to influence sentencing, and how it can be manipulated to incite genocide. You will find out how, when, and why things that we find disgusting can also be very amusing. You may be tickled or shocked to learn that depending on where you live, the same depictions of porn are either vomitously repulsive or erotic entertainment. You will learn what things are especially disgusting to you and how generally stoic or squeamish you are compared to others.

Disgust teaches us about the inner workings of our brains and personality. It illuminates why we have certain reactions to groups of people and why we treat various individuals the way we do, for both better and worse. Disgust informs us about our culture, our food preferences, our sexual passions, our laws, and our customs. Most importantly, disgust reveals the fundamental concerns that underlie our existence. Indeed, by deciphering disgust we come to more fully understand what it means to be human. I hope you enjoy the journey as we unravel all *That's Disgusting*.

THAT'S DISGUSTING

Chapter 1

LET'S EAT

The alarm sang Its 7 a.m. tunc, and Akiko awoke feeling refreshed but hungry. She'd had a tiny dinner the night before and was ravenous. Moments later Akiko was in the kitchen, opening the refrigerator and pulling out a carton of nattō. "Mmm, this will be good," she murmured as she eagerly dug her chopsticks into the sticky, slippery beans.

At the same time, on the other side of the globe, Mia was opening a bottle of Cabernet and placing goat cheese-filled olives, fresh Stilton, and taleggio on a platter with grapes and crackers for the guests that would arrive soon. She hadn't eaten since lunch and the luscious cheesy treats were oh-so-tempting. She checked the clock again, sighed, and then succumbing to pleasure sliced off a soft, gooey wedge of taleggio.

Nattō is a stringy (the strings can stretch up to four feet), sticky, slimy, chunky, fermented soybean dish that the Japanese love and regularly eat for breakfast. It can be eaten straight up, but is usually served cold over rice and seasoned with soy sauce, mustard, or wasabi and can also be garnished with green onion, fish flakes, raw eggs, or radish. The latest figures show that over 14 billion pounds of nattō are produced annually for Japan's population of 127.9 million people.[1] Aside from its alien and complex texture, natto suffers from another problem—odor—at least for Westerners. To me, natto smells like the marriage of ammonia and a tire fire. Although this might not be the worst smell combination ever, it has zero food connotation, and there is not a Westerner I know of who at first attempt can get nattō into their mouth.

And then there's cheese—considered by Westerners to be anything from a comfort food to a luxurious delicacy. Its texture can range from hard to liquidy—with all variations in between. A good chambré'd taleggio, gorgonzola, or Brie might be described as sweatily slimy. Cheese also has its fair share of aromatic obstacles and, depending on the circumstances, may be confused with vomit, stinky feet, or a garbage spill. Taleggio is ranked as one of the worst-smelling cheeses in the world even by cheese connoisseurs.[2] Many Asians, however, regard all cheese—from processed American slices to Stilton—as utterly disgusting and the literal equivalent of cow excrement, which considering that it is the rotted consequence of an ungulate body fluid, is technically correct. Yet both nattō and cheese are nutritious and high in protein, vitamins, and calcium. Both foods are also rotted, or to put it more politely, "fermented." Akiko and Mia are enjoying two well-known local favorites, and it turns out that every culture has a favorite fermented food. Controlled rot tastes good.

Controlled rot—fermentation—is the process of converting carbohydrates into alcohols and carbon dioxide with the help of microorganisms like yeasts and bacteria, typically in an anaerobic—that is, oxygen-depleted—environment. Disgusting as this may sound, many of the fermented products that we enjoy today, such as bread, wine, beer, and cheese, were conceived because some benign bacteria serendipitously fell off human skin and landed in a pail of milk, for example.[3] In fact, the same bacteria responsible for foot odor are in many cheeses. Even though fermentation involves the introduction of foreign microbes into the things we eat, these microorganisms actually protect against the invasion of dangerous germs. Of the millions of bacterial species on the planet, only about fifty are harmful to humans, and we are generally protected against these by the throng of safe bacteria that live in our bodies, the soil, water, and air.

Fermentation does not require human intervention. Many foods will naturally ferment once they are past their prime, but humans have been meddling in the process for the benefits of alcohol for an estimated 8,000 years. Fermentation can also be manipulated to encourage mold growth, as in the production of blue cheeses and the Chinese delicacy of "hundred-year-old" eggs. The common flavor characteristics of fermented foods tend to be strong aromas and sour taste, due to the acids that are a by-product of the fermentation process.

SOMETHING IS ROTTEN IN THE STATE OF . . .

A quick jaunt across the globe for some favorite ferments will lead us to kimchee in Korea, which is fermented vegetables (usually cabbage); *gravlax*, the fermented raw salmon enjoyed in Norway; *injera* in Ethiopia, is the spongy, fermented flatbread used as an edible base for picking up various stews and raw minced meat and vegetables with your (right) hand; *chorizo* in Spain, which is fermented and cured uncooked pork sausage; and the many forms of fermented dairy that are adored and consumed from India to Indiana.

Among the most hard-core variants of fermented food is the Icelandic delicacy *hákarl*. Hákarl is made from the Greenland shark, which is indigenous to the frigid waters of Iceland. It is traditionally prepared by beheading and gutting the shark and then burying the carcass in a shallow pit covered with gravelly sand. The shark needs to be buried on an embankment because with the carcass on an incline, the body fluids can more easily seep out as the shark rots while weighted down with more sand, gravel, and stones. The shark corpse is then left to decompose in its silty grave for between two and five months—the amount of time depends on the season, milder weather entails less time. Once the shark is removed from its lair, the flesh is cut into strips and hung to dry for several more months. During the drying process a brown crust develops over the meat which is removed prior to final preparation. Hákarl has a pungent, ammoniac, fishy odor which causes most newbies to gag or vomit. Not for the faint-hearted, it is typically consumed with shots of *aquavit* and is heralded as possessing powers to make one stronger. An extremely acquired taste, hákarl was described by the globe-trekking celebrity chef Anthony Bourdain as "the single worst, most disgusting, and terrible tasting thing" he had ever eaten.[4]

Necessity is the mother of invention, and there is nearly nothing so necessary as food. The "invention" of hákarl arose because eating the meat of this shark when fresh is poisonous due to the high levels of uremic acid in its flesh. It turns out that the solution is to let the meat rot because fermentation causes the uremic acid to evaporate—but there is a serious risk of botulism if this isn't done properly. Traditional methods of fermenting foods in the ground are actually less risky than modern methods involving storage in plastic containers. Northern native populations who ferment fish and animals have seen an increase in botulism cases since the newer, easier methods of fermenting have been introduced. The plastic containers create airtight conditions that the botulinum bacteria thrive in. In addition to making a toxic food edible, fermentation creates new flavors and foods, preserves food for extended periods of time without refrigeration, and decreases cooking time.

At an international convention of food challenges, you might try to wash down your hákarl with the Ecuadoran aperitif *chicha*, which combines the alcoholic perks of fermentation with a disgusting bodily fluid. Chicha is made from a masticated blend of boiled maize (or yucca root) and human saliva. One of my students vividly told me about her personal adventure with chicha while she was living in Riobamba, Ecuador.

> I was visiting the hut-home of my host family's maid, sharing with
> them what it was like to go to school in the US, when I noticed a

ritual taking place a few feet from me. While continuing their con-
versation with me, the women, including my host, began lifting
what looked like corn flour to their mouths, chewing for a while,
and then spitting the concoction into a milk jug. The spittle–corn
combination vaguely resembled the vomit of an infant as it spilled
from their lips to splash together inside the jug. But no one bat-
ted a lash. Clearly, for everyone but me, this was business as usual.
When the dry corn flour was gone the milk jug was capped and the
eldest in the room, a woman of at least eighty, stood and brought
my attention to a loosened area of soil. The old woman then bent
down and dug out a twin jug from the ground whose contents
were similar to the one they had just filled but distinctly darker
and more pungent-smelling. The jug that I had just watched them
fill was then buried so that, I later learned, the mixture would ferment. The concoction which had just been unearthed was then
poured into a gourd bowl and everyone was visibly excited for
me to partake in the tradition of passing it around. I was offered
what seemed like the enviable position of drinking second—after
the eldest had. But at that moment I wished I was anywhere else
but there. The acrid-smelling soup was handed to me and I gave
myself a private pep talk before managing to take a quick gulp
and passing the bowl to the woman beside me. As I sat there, my
mouth full of chicha, I had to force myself not to gag and spew the
contents of my mouth out over everyone. Finally swallowing the
stuff, I can only describe this brew as being like warm, thick, lumpy,
vinegary beer.

My favorite fermented challenge, because I'm a cheese lover but
am mortally repulsed by worms, is *casu marzu*. Casu marzu is a sheep
cheese popular on the Italian island of Sardinia. The name means "rotten cheese" or, as it is known colloquially, "maggot cheese," since it
is literally riddled with live insect larva. To make maggot cheese you
start with a slab of local sheep cheese *pecorino sardo*, but then you let
the cheese go beyond normal fermentation to a stage most would con-

sider infested decomposition—because it is. The larvae of the cheese fly (*Piophila casei*) are added to the fermenting cheese and the acid from their digestive systems breaks down the cheese's fats, making the final decomposed product very soft and liquidy. By the time it is ready for consumption, a typical casu marzu contains thousands of larvae. Because locals actually consider it *unsafe* to eat casu marzu when the larvae have died, casu marzu is served with the live larvae actively squiggling.

The larvae are translucent white worms—maggots—about one-third of an inch long. Some people clear the maggots from the cheese before consuming it; others do not. Those who eat it with the maggots still milling about, cover the cheese with their hands to prevent the maggots from leaping onto them or anything else, since when they are disturbed the maggots can jump distances of up to six inches. The cheese is typically complemented with the Sardinian bread *pane carasau* and Cannonau, a strong red wine. If you think this is the only cheese now going on your "do not eat list," be aware that other regions in Europe cultivate cheeses with live arthropods, such as the German Milbenkäse and the French Mimolette, both of which rely on cheese mites for aging and flavoring.

It is no accident that you likely feel revolted by my descriptions of many of these fermented comestibles. The most basic form and elemental purpose of the emotion of disgust is to engender an avoidance of rotted and toxic food. So why is fermented saliva, decomposed shark, and maggot-ridden cheese so desirable? Is it just a quirky paradox of the human condition that we eagerly consume that which gives off all the signals of putrefaction? No. It illustrates, that to a very large measure what is disgusting, or not, is in the mind of the beholder.

ONE MAN'S MEAT IS ANOTHER MAN'S POISON

The primary way that we come to have a position or belief about something is through cultural learning. Your parents taught you to eat specific foods using certain specialized utensils: a spoon for ice cream, for example. In just the same way, we learn which foods are disgusting and which are not through our cultural heritage. "This is on the serving platter, so this is what we eat." One reason why foods are so locally meaningful is because they come from the flora, fauna, and microbes of a given region, which can vary markedly between geographic areas.[5] The bacteria involved in making *kimchee* are not the same as those used to make Roquefort. Another reason why culture is such an important determinant of the meaning of food is because we use food as a way of establishing brethren or foes, and as a mode of ethnic distinction. "I eat this thing and you don't. I am from here, and you are from there." Not only is the meaning of food learned through culture, food is used to establish cultural boundaries and borders.

Prohibitions against consuming specific foods have a long history of being used to keep the "wrong" people away. For example, under the Roman Empire, rabbis were concerned about the rise in Gentile–Jew fraternizing, especially under the influence of alcohol and the slippery slope of intermarriage it portended, and so forbade the consumption of Gentile wine, beer, and food. The Talmudic scripture puts it succinctly: "Their bread and oil were forbidden on account of their wine, their wine on account of their daughters, and their daughters on account of 'another thing.'"[6]

Wine in particular was blacklisted, because in addition to the recklessness that comes with overindulgence, wine is used in specific ways during religious rituals that vary among Jews, Christians, and Muslims. Therefore, careless use of wine and wine-based products like vinegar (because it was made from Gentile wine), and even foods pickled in vinegar, was feared to lead to idolatry. The mystical essence of the winemaker was believed to enter into his brew, and so by consuming his wine you could became spiritually contaminated and converted to his religion.[7]

Another important signal for cultural distinction is a consequence of food consumption. You smell like what you eat. If you eat a lot of garlic then your body odor will have a garlicky redolence. This is because the odorous compounds in the foods we eat are radiated through our skin and sweat. These aromas are both sociological markers for a shared food culture—"I like how you smell because it means that you and I eat the same thing"—but unfortunately also have a long and tragic history as a means to ostracize groups and promote prejudice. Jacques Chirac, the former prime minister of France, infamously empathized with the French worker for "having to put up with the noise and smell of the immigrant family living off welfare next door." The foreigners eat strange meals that have strange aromas and their bodies reek of their strange food. These unfamiliar aromas are associated with the unwanted invasion of the foreigners and thus are unwelcome and repugnant. Because of the uniquely potent link between smell and emotion, the visceral disgust a scent can provoke cannot be easily overcome, and the foreigners and their food become a stench to be eradicated.

Conversely, a person can become more accepted by eating the right foods—not only because your body odor will no longer smell unfamiliar and "unpleasant," but because acceptance of food implies acceptance of the larger system of cultural values at hand. This is why it is important to graciously eat the meal offered by your host from a foreign land, especially if your host has any power over you. When visiting the Chinese dignitary to convince him of your environmental business plan, you will have to consume the bull penis soup. The name is self-explanatory. It might even make you stronger. The Chinese Olympic team consumes a bowl of bull penis soup every day.[8]

FOOD FOR THOUGHT

What makes bull penis soup, chicha, or casu marzu delectable? In addition to teaching culinary traditions, cultural learning imparts knowledge about what is safe to eat. If after watching your friend drink a bowl of saliva and corn mash you see her happily working in the field, you're more likely to drink chicha the next time it's offered, but if you see her vomiting after drinking it you are much less likely to. The influence of our peers, and especially knowledgeable peers, guides us where other cues, such as signs of putrefaction, may be evident but misleading. To use cheese again as the example: on visual inspection, melted cheese is the pinnacle of shiny, oozy, off-white yellowishness and should be disgusting, if disgust helps us avoid what is decomposing, rancid, contaminated, or otherwise sickness-inducing. However, watching Adam Richman, host of the fabulous and fattening—just by watching—television show *Man v. Food*, who has "held nearly every job in the food industry," glory in oodles of melted cheese on a regular basis teaches me that this mucusy-looking gooeyness is good. Maybe if I watched Adam eat casu marzu I could be persuaded too. We learn both the meaning and safety of food from our peers, and this learning shapes our psychological interpretation of food.

Our psychological interpretation of food is even capable of transforming water into a hideous drink, as W. C. Fields famously put it: "I never drink water because of the disgusting things that fish do in it." Personally, I have to stop myself from thinking of shrimp as wood lice, because otherwise I wouldn't be able to make it through my next wedding buffet. In fact, our favorite shellfish are in the same subphylum as these insects: crustacea. The latest phylogenetic evidence actually indicates that all insects are crustaceans, just as birds are modern-day dinosaurs.[9]

Other shellfish meals have also oscillated between torment and treat. Lobsters, the subject of seasonal culinary festivals and a sumptu-

Headless shrimp, side view   Wood louse, side view   Wood louse, top view

Figure 1.1

Shrimp and Wood Louse Comparison

ous indulgence, were considered aquatic vermin when they were discovered by European colonists in the 1600s. This sea pestilence, which littered the shores of New England, was used as fish bait and fertilizer and fed to the poor, orphans, slaves, and prisoners. Excessive serving of this "junk" food led slaves in Massachusetts to rebel and have mandated into their contracts that they would eat lobster no more than three days per week.[10] Later, Massachusetts passed a law forbidding dishing up lobster to prisoners and servants more than twice a week—a daily lobster dinner was deemed cruel and unusual punishment.[11]

One of the more difficult edibles to get one's head around is human placenta. The placenta is a vascular organ that develops in the uterus during pregnancy and connects the fetus to the mother's circulatory system; through it the fetus gets nutrients, eliminates waste, and exchanges oxygen and carbon dioxide. The fetus is attached to the placenta via the umbilical cord, which is cut immediately after birth and the placenta is discarded as hospital waste. For most of us, the thought of eating placenta after it has spilled out of the mother's birth canal is the furthest thing from appetizing. Nevertheless, most female mammals consume their own placenta after giving birth, and some human cultures—indeed, individuals within our own culture—revere its consumption.

Many groups place spiritual value on the human placenta and prescribe rituals for its use. The Navajo bury a newborn's placenta within the sacred Four Corners of the tribe's reservation so as to bind the infant to its ancestral land and people.[12] Among the cultures who directly advocate consuming human placenta are the Chinese and Vietnamese. Traditionally, Chinese nursing mothers were advised to boil their placenta and drink the broth to improve their milk quality. The Chinese also use dried ground human placenta in various medicines. Eating human placenta is even alleged by some Western health-care practitioners to reduce postpartum depression.[13] Purported to have exceptional protein and nutritional properties, placenta is also an ingredient in expensive French cosmetics that promise beautiful, youthful skin. Some bold and eccentric Westerners even promote placental gastronomy. In 1998, a British cooking show called *TV Dinners* featured a London couple who, to celebrate the birth of their granddaughter, made a pâté out of her placenta and served it to family and friends on focaccia bread. Many viewers filed complaints, and the Broadcasting Standards Commission concluded that the program had breached a taboo and that this meal should not have been aired. Besides the fact that placenta is organ meat, which we typically regard as the least appealing body part to eat, the more troubling question that emerges is whether eating placenta constitutes cannibalism. After all, it is *human* organ meat that one is eating.

THE ULTIMATE FOOD TABOO

On Friday, October 13, 1972, a chartered flight with forty-five people on board, comprising the Old Christians Rugby Club, their friends, family, and four crew members, took off from Carrasco International Airport in Uruguay on route to Santiago, Chile, for an exhibition rugby game. As Uruguayan Air Force flight 571 flew over the Andes nearing the end of its journey, the jet crashed into a snow-packed glacier in a remote mountainous region on the border of Chile and Argentina. Twelve people died instantly, including the entire flight crew, and by the next morning five more were dead. The sister of one player succumbed to her injuries on the eighth day of the ordeal. A week later another eight, including the last woman alive, were killed by an avalanche that buried the fuselage that the survivors had been using as a shelter. Two days before Christmas, another three now dead, the last sixteen survivors were rescued by military helicopters.[14] How did these castaways survive for seventy-one days in snowbound, subzero conditions without any provisions?

By the end of the first week, anything that could count as food—some chocolate bars, jam, and a few other snacks that were retrieved from the wreckage—had been consumed. The stranded were all also severely dehydrated, but after the first few days had discovered a way to melt enough snow to stave off a deadly thirst. In the high altitude and cold they were burning far more calories by simply breathing than they would have in warm, sea-level conditions and the group was quickly entering the realm of serious starvation. In various last-ditch efforts, they foraged for anything that might be edible. They chewed leather from the luggage and tore up the plane seats in search of straw, only to find upholstery foam. They scrabbled over the ice-encrusted tundra, but no flora or fauna, not even a single insect, could be found. Then, on the tenth day, the scent of blood from one of the injured passenger's wounds catalyzed the impulse that was lurking just beneath the surface, and Nando Parrado confided to one of his teammates, Carlitos Páez, the unspeakable thought that had been on his mind. The next morning they held a group meeting.

Nando and Carlitos told the others that there was no way any of them would survive without eating, and that the only thing they could eat was the protein-rich flesh of their dead kith and kin. They intuitively knew to use the rationale that how one thinks about food greatly influences whether one will be able to eat it, and encouraged the others to think of their friends as "only meat." They also appealed to the survivors' sense of justice, arguing that if the bodies of their compatriots could keep the rest of them alive then their deaths would not have been in vain. In other words, they tried to make the act of cannibalism morally justifiable. The survivors were all Catholic and the staunchly devout among them initially refused to eat their dead brethren; however, they all later succumbed to their drive to survive. They also made a pact; if any of them should die, the rest had their permission to use their body as food.

The "cannibals" had to be very careful in rationing their newly found food, as they had no idea how long they would be stranded, so they ate very little and used every edible portion of the bodies that they could, from skin to intestines. To make the "meat" more palatable they cut it into small pieces and let it dry in the sun. On rare occasions they were able to make a fire and cook it, which apparently "improved its taste dramatically." Finally, after a daring and nearly impossible trek down the glacier by Nando Parrado and a teammate, Roberto Canessa, they were discovered and a military rescue operation saved the remaining men.

The story of the horrendous ordeal and amazing survival of these young rugby players in the Andes instantly became sensationalized and has been the subject of two books, two films, and numerous television documentaries. There have been other stunning stories of survival against all odds in the years since, but none has so enduringly captured the world's attention as this one. The reason we are so fascinated by this story is because of the method of their survival. One of the most unspeakable taboos was broken, and it was sanctioned. Under certain circumstances cannibalism can be accepted.

Indeed, up until very recently, cannibalism was regularly practiced by the Biami tribe in Papua New Guinea, a remote island north of Australia. The last officially recorded cannibalistic killing in Papua New Guinea occurred in 1993, but the practice may go on today. The arrival of Christian missionaries in the 1960s—though they are depicted in old Hollywood movies being boiled in pots by savages ready to eat them—was actually what ended tribal cannibalism, and according to local tribal elders no white men were ever eaten. The main reasons given for why cannibalism took place were revenge for a wrongful death and getting rid of suspected sorcerers, but ultimately it was about supplementing the Biami's very sparse diet with protein, and there is some indication that this was at times the main objective—eating humans simply for their meat because there was no other nutritionally rich food source. In fact, the tribal members who took part in eating human flesh did not consider it any different than the rare pig meat they would also sometimes consume, and, according to them, we and pigs pretty much taste the same.[15] Necessity is the mother of culinary invention, and I believe that prohibition against murder, which is a major sin in most religions, arose in part as a method to stop us from hunting and eating one another.

RIGHT AND WRONG FOODS

The cultural divisions on disgust or delicacy aren't confined to live, human, or rotted categories. There are many fresh, thoroughly cooked, vegetarian, and commercially processed comestibles from other cultures that we find repulsive; this is due to what we think food *is* or *should be*. Frog congee (rice porridge) is Chinese comfort food, but I'd much rather have macaroni and cheese. This is only because I did not grow up with Mama making frog porridge. We don't even need to travel far to discover major cultural disparities. Wintergreen mint is rated among the most preferred flavors in North America, but in Britain this same mint is considered revolting. This is because in the US we only know wintergreen as the sweet flavor of a candy or gum treat, whereas in the United Kingdom wintergreen is used to scent toilet-cleaning products and in medicinal balms.[16]

Foods that we individually like a lot, are very familiar with, and in themselves pose no challenge, are considered repulsive if they are mixed in the "wrong" combinations. Meat mixed with ice cream, for example, may sound repulsive. But "wrong" is in the mouth and mind of the eater. The Japanese love their ice cream—and although we may think we're being adventurous when we have a cup of fried green tea ice cream after our sushi dinner, how about some octopus, ox-tongue, or chicken-wing ice cream? Just about anything you can imagine—including charcoal, tulips, and beer—has been concocted into sweet and creamy ice cream in Japan.

We also consider eating condiments by themselves as highly inappropriate and disgust-provoking. Watching someone sit down to a jar of mayonnaise or chomping on a stick of butter would be revolting to most civilized diners. Why? Because of the animalness of the act. A pig would happily gulp down a stick of butter, but we must restrain ourselves from behaving animalistically. We have rules and codes for eating so that we can maintain a barrier between our civilized humanity and the wildness of beasts.

Codes for eating dictate what is appropriate to eat as well as how we should eat it, but they vary tremendously between cultures. Noises made, or not made, while dining are one example. In the Middle East, belching at the table after a meal is a great compliment, but in North America, it is vulgar and disgusting. Chewing with your mouth open and making slurping sounds while eating in China is entirely accept-able, but Americans teach their children never to do it. Whether you hold your fork in your right or left hand, bend your knife and fork over your plate, or hold your utensils vertically is either appropriate or boor-ish depending on whether you are in the UK or the US. Indeed, if you accept food or eat with your left hand rather than your right, it could expel you from the table or horrify your hosts in many African and Asian countries.

Within a very short time span my own North American culture has witnessed a change in attitude toward a "disgusting animalistic eating" display. I grew up learning that chewing gum, like a cow chewing its cud, was unseemly and not to be done outside the company of my school friends, and that I should especially not chew open-mouthed or while talking or eating. But now, chewing gum is something that both chil-dren and adults do alike in all kinds of social settings. I recently gave a keynote lecture on the psychology of aroma and flavor at the Interna-tional Chewing Gum Association annual meeting held in a posh hotel in New York City, and, not surprisingly, everyone was chewing gum. But what stunned me was that the majority of the audience were suits in their fifties, and they masticated audibly and continuously through all the talks, while they asked me questions, and even as they snacked on pastries during the coffee break. I had never seen this in a business context nor with anyone over the age of twenty-five. Curious, I com-mented to a few of them "how much gum-chewing rules seemed to have changed." They agreed and blithely chomped on.

Food rule-breaking in the form of consuming excessive quantities of unhealthy high-calorie foods elicits repulsion at our animalistic glut-tony, but it also holds a strange attraction for us. In the summer of 2010, the Indiana state fair experimented with a variety of artery-clogging food novelties, including an 800-calorie "donut burger"—a beef patty covered in melted cheese and sandwiched between two glazed donuts (a Big Mac has a mere 540 calories). Vendors also served up fried Oreos and deep-fried butter, and another creative treat, "pigs in the mud"—chocolate-covered bacon—sold by a local farmer, who explained that any marketing scheme to promote his livestock was a good one (I can personally attest that this last concoction is delicious). The Indiana state fair fare was covered on National Public Radio, and a few days afterward, the commentators read listener e-mails in response to the report, which ranged from a man in Pennsylvania who denounced the food as being "gross perversions" and "gastronomic pornography" to a woman in Southern California who, upon hearing about the donut burger, abandoned her healthy dinner of wild salmon and veggies to order a pizza.[17]

Why are we offended and disgusted by someone licking a gob of mayo off their finger or eating a donut burger for dinner? The reason is because this type of gluttony is animalistic and eating animalistically makes us animal-like. We are what we eat, and we are how we eat. Moral vegetarians—people who become vegetarians because they are against animal cruelty or for ecological reasons—think that people who eat meat are more aggressive and animal-like than their fellow leaf-eaters,[18] and college students, vegetarian or not, judge people who are described as eating "unhealthy" high-fat diets as being less kind and considerate than people who are described as eating a "healthy" vegetarian diet.[19] It is our thoughts—our psychology—that make food and the people who eat that food disgusting, or not.

THE TROUBLE WITH THINKING

Psychologically framing something edible in "food" terms can help us consume what we would ordinarily find impossible to eat. On the flip side, thinking in certain ways about what we are eating can be a deterrent even when we want to eat it. If you start musing about *Charlotte's Web*, you might start to have a problem chowing down on those finger-licking baby back ribs.

I was recently the victim of too much of my own thought. I was out for dinner at an expensive, pretentious locavore restaurant and was intrigued by a special appetizer of the night, "duck oysters," which were then further described as "duck fries." Given my love of French fries and liking for duck and oysters, I thought this sounded like a tasty option. But when my dish arrived and I saw what looked like breaded calamari only in the shape of small bulbs, I was taken aback. I took a little bite.

"What do you think this is?" I asked someone else at the table.

"Duck testicles, obviously," she replied. "You know, Rocky Mountain oysters[20]—when it says oysters it means testicles, unless of course it's really oysters." She laughed.

"Oh," I said, not amused. And that was it. I don't know why it never occurred to me that "oyster" was that notorious euphemism, but as soon as I started to think about what those little white balls were, and even though I had tasted them and found them "not bad," I could not go on and instead gave my dish to my husband who happily received it, as his food motto always is "If it tastes good, eat it."

My husband notwithstanding, North Americans are prudes about many foods, and we especially don't like the idea of eating anything that is still kicking. In Korea, you can adventurously order *sannakji*—a whole, live octopus, although choking is a serious hazard, as the tentacles stick to everything, including your throat. In China, chefs can serve you monkey brains from a living monkey sitting at your feet with its skull carved open, and even a fish that has somehow been kept alive through the deep-frying process. By contrast, you are unlikely to find a live dinner in most places in the West, apart from the trendiest big-city restaurants.

One reason why eating animals while they are still living is such a difficulty is because it brings us up close to our fears of our own livingness, and forces us to confront the fact that we are mortal creatures with ticking time-bomb life spans, and so we shield ourselves from this reality by hiding behind the convenience of modern-day grocery shopping. Most of us purchase and consume food that has lost all resemblance to the animal that produced it. The meat we see at the grocery store and even in most gourmet butcher shops is headless, tailless, bled, and carved into neat symmetrical portions. The words we use to describe what we eat further distance us from its animality. What is a "beef," or for that matter a "Big Mac"? Would there be more vegetarians among us if we had to catch, kill, and carve up a cow in order to have a Big Mac—that is, face our dinner eye to eye? On that note, many readers would agree that the most grotesque animal part to consider eating is an eyeball, yet for Inuit children in northern Canada seal eyeballs are a special treat. In Boston, there might be more vegetarians if we had to look into the eyes of our would-be dinner, but not in Inuvik. Culture defines what is disgusting or not to eat.

It may come as a surprise, but vegetarians are not more disgust-sensitive than carnivores. In fact, the reverse is true. Meat eaters are *more* sensitive to disgust than vegetarians, and how well-done they like their meat is a marker for just how disgust-sensitive they are. People who like their steak to be indistinguishable from the sole of an old shoe are especially squeamish, while those who like their steak "blue" are a lot less so. However, some vegetarians are more disgusted by meat than others. In particular, people who become vegetarians for moral reasons typically end up finding that the meats they used to love now disgust them. By contrast, people who become vegetarians for health reasons still appreciate the pleasure that savoring a juicy filet mignon can bring.[21] What makes moral vegetarians become disgusted by meat? It turns out that thinking makes it so.

It has been found that just *thinking* of eating meat as immoral causes it to become disgusting. In one experiment, Daniel Fessler, a prolific evolutionary psychologist and disgust researcher, along with his colleagues at UCLA, explored the connection between disgust sensitivity and meat consumption using a Web-based survey. Over one thousand respondents who ranged in age from thirteen to seventy-nine were polled, and the responses showed that it isn't inherent squeamishness that predisposes any of us to become a vegetarian for moral reasons. Rather, it is how we construe the meaning of meat as food that makes it immoral (or not), and its immorality then makes it revolting.[22] If we feel that eating meat is *evil*, whether for environmental, religious, or humanitarian reasons, we come to view meat itself as disgusting, but if we simply don't eat meat because our doctor told us we should stop, then we still long for a juicy steak now and then. As Robert Louis Stevenson said: "Nothing more strongly arouses our disgust than cannibalism, yet we [meat eaters] make the same impression on Buddhists and vegetarians, for we feed on babies, though not our own."

We are most comfortable eating food that we feel emotionally neutral toward—whatever that may be. Omnivores in the West are horrified by the thought of eating the animals we bring into our homes and love—our pets—so dogs and cats are out. And we don't like the thought of eating anything we feel "disgusted" by—so lizards and snakes don't get munched on much either. Rather, chickens and fish seem benign, unlike us, and generally unrepulsive, and so we have the easiest time consuming them. This is also why a large subset of "vegetarians" still eat some animal protein, typically fish and eggs. In the hierarchy of avoiding meat, mammal meat is rejected first. The underlying justification combines health concerns—mammal meats like steak and bacon are high in fat and therefore unhealthy—with emotional concerns—it is more inhumane to slaughter mammals because we are mammals too and as such we confer a higher degree of kinship and consciousness on cattle and pigs than on proto-dinosaurs (birds) or cold-blooded fish. Given that many vegetarians allow themselves animal protein so long as the donating animal isn't perceived as being very sentient, could they, or the rest of us, be persuaded to try another form of non-mammal protein?

BUGS ARE NUTRITIOUS AND CAN ALSO BE DELICIOUS

In the middle of a three-hour seminar class I teach at Brown University called "The Psychology of Aversion" I typically give a short break, and because the students complain to me that they get hungry in the middle of the afternoon I pass around mini chocolate bars. About half way through the semester in the winter of 2010, one of my students asked if he could share something of sustenance with the students. Frasier, the food-offering student, didn't tell me what he would be bringing in, but something in his shy smile made me agree. At the designated break, Frasier dug a crumpled paper cup out of his knapsack and, turning pink in the face, announced that he had grilled crickets over rice on offer. He explained that in another class on environment and sustainability a guest speaker who sells bugs to restaurants and for pet food had come to explain and encourage the eating of insects. The bug chef had grilled crickets with olive oil, salt, and pepper and served them over rice for the class. He had also filleted and grilled the shoulder meat from the apparently tender and giant Thai beetle and prepared cicadas as well, but Frasier had saved only the grilled crickets.

"Who wants to try one?" he asked nervously as he pulled a box of Triscuits from his knapsack, commenting that the crackers were a good accompaniment. To my surprise, about two-thirds of the students immediately accepted the cricket–Triscuit hors d'oeuvre. I was actually among the abstainers. The consensus was that it was "not so bad"—that the texture was the surprising part, crunchy and then squishy, but apart from that it tasted "pretty good," and that the Triscuit cracker helped. Many also said that the rice on the bottom of the cup was a turnoff, but understood why they were brought in this way—the bugs were easier to pick up off rice than if they had been nakedly intertwined at the bottom of the paper cup.

What about the refuseniks? The few who didn't venture said that they didn't want to try it because it "just looked too gross" or that the thought of eating insects was too repulsive. But why did I rebuff the insect snack? I truly wanted to be among the brave and eat one, but empowered as professor I didn't feel a need to disgust myself. The grilled crickets with their bent bodies and spindly antenna thatched around one another sitting atop the grayish rice were really repulsive to me, and as with the "duck oysters," my thoughts turned my stomach. But it seems that I may have to get over my buggy aversion soon.

GREEN AND EARTHY EATING

In 2009, Vij, an upscale Indian restaurant in Vancouver, Canada, had a line of eager diners outside its door from opening to closing seven days a week. What they were all waiting for was the famous "cricket paratha," a twist on the traditional Indian flatbread, which rather than being made from whole wheat flour is made from crickets. It takes about 350 crickets to make enough paratha for two people. Moreover, unlike its carbohydrate cousin, cricket paratha has as much protein as a steak, but is three times higher in iron and much lower in calories, fat, and cholesterol. With the right spin and artful camouflage, eating wriggling, squiggling, squeaking, insects is in vogue and getting ever more popular. Not only could buggy dining become chic, it could also become virtuous.

Moralization is a process whereby an object or activity is transformed from morally neutral to morally charged—usually in the negative—because of a newly invoked moral implication.[23] The change in attitudes toward cigarette smoking in North America over the past fifty years from cool to debased is a classic example. Spearheaded by mass media health campaigns that warned how "secondhand smoke" could cause serious harm to innocent bystanders, especially children, the insinuation was that smokers were committing the near equivalent of negligent homicide and were thus evil people.[24] Moralization can also work in the reverse and may be the strategy that will be used to turn the culinary consumption of bugs from nauseating to noble. I can imagine billboards of the future featuring slogans like "Every cricket dinner is a gift to mother earth." Green-oriented advertising has succeeding in making many sneer at gas-guzzling Hummers and instead aspire to drive Chevy Volts. Could a similar marketing persuasion make us crave a plate of bugs?

Raising livestock has an enormous and negative environmental footprint. Livestock agriculture requires vast amounts of land and feed and produces more of the greenhouse emissions that cause global warming than cars, planes, and all other forms of transportation combined. Ranching is also a major cause of deforestation worldwide, and half of all fresh water on the planet is used for livestock. It takes an astounding 1,048 gallons of water to produce one gallon of milk, and the water necessary for one pound of steak equals the amount of water consumed by a family of four for a year—approximately 208,000 gallons.[25] A less staggering but still shocking statistic is that it takes fourteen pounds of feed to produce one pound of beef. By contrast, it only takes two pounds of feed to produce one pound of cricket meat. Crickets will eat just about anything, and they consume minuscule amounts of water. Moreover, harvesting insects emits a fraction of the greenhouse gases, like methane, that are belched out by livestock. Insect husbandry can also be done in small spaces and therefore drastically reduces the amount of land needed. Insects could even solve the projected world food shortage.

The United Nations forecasts that the world's population will surpass nine billion by the year 2050—more than triple the number of people that were on the planet in 1950.[26] Westerners consume around 250 pounds of beef and pork combined per person annually. Asian and developing nations consume less, but with increasing modernization and population growth one could conservatively imagine that by 2050 at least five billion people will expect to eat their share of 200 pounds of meat per year. In order to feed this many people this much meat, the world would need to produce fourteen billion pounds of beef and pork annually, and there is not enough land or nutrients on earth to support this kind of production. If entomophagy (insects as food) became popular, a way to end world hunger would be within reach.

Thailand is the current world leader in insect farming, with about 15,000 farms raising locusts, grasshoppers, and mantises for human consumption. Insects also feature in the diets of rural Laos, Vietnam, Colombia, Brazil, and Mexico. Among the most popular dishes are deep-fried crickets, barbecued larvae, and grilled tarantulas. Without leaving the US, the adventurous gourmet can meet up with Gene Rurka, an ecologically-minded exotic-food chef and farmer from Somerset, New Jersey. Among the dishes he is famous for are teriyaki Madagascar hissing cockroaches, "wormzels" (baked worms that sizzle into pretzel shapes), banana canapés topped with maggot pupae, and tarantula pops.[27] Rurka served up these items and more, including baked scorpions atop a slice of cucumber with herb cream cheese, when creating the menu for the Explorers Club's hundredth birthday party—a black-tie affair held at the Waldorf Astoria in 2004 with an 1,800-person guest list.[28] Indeed, Rurka concocts the buffet of "exotic" hors d'oeuvres that precedes the main meal at the Explorers Club's yearly gala dinner, where in addition to cockroaches and spiders, members can sample rattlesnake, beaver, and sweet and sour bovine penis, to name but a few crowd-pleasers. There is also a cocktail station which has variously featured a martini with a goat, lamb, or calf eyeball garnish (stuffed with an olive or onion too).[29]

If reading about bug grub has somehow made you hungry, you can learn how to prepare an assortment of delights, from appetizers to desserts, in cookbooks such as David George Gordon's *Eat A Bug Cookbook*, or Dr. Julieta Ramos-Elorduy's *Creepy Crawly Cuisine: The Gourmet Guide to Edible Insects*.[30] On the other hand, if what I've described sounds like an episode of *Fear Factor* to you, you may wonder how it is that people came to eat an assortment of horrifying foods, from wriggling insects to putrefying meat. The answer again is necessity. Like cannibalism and the "invention" of hákarl, starvation is an enormous motivation when it comes to figuring out how to turn possible nutrient sources into edible foodstuffs.

A small saving grace, however, is that food tastes better when you're famished. A number of studies have shown that how good food tastes depends on how hungry or full you are. We innately like the tastes of sweet and salty and dislike bitter. When you're starving, your taste buds become more sensitive to sweet and salty tastes, but bitter taste sensitivity doesn't change. Sweet signals carbohydrates and salt signals proteins, and you need to consume these nutrients when you're starving. It seems that nature has helped us in this predicament, as anything from human flesh to cockroaches will taste sweeter and saltier when we're really hungry, but not more bitter.[31]

PIGGING OUT

Even though chocoholics will confess that chocolate becomes unpleasant when they're surfeited,[32] there are some people who, beyond merely feeling full, will try to eat as much of a favorite snack food as their stomachs can physically manage. We've all gone a little crazy at the all-you-can-eat buffet, but have you ever stuffed yourself with eleven pounds of cheesecake or over one hundred hamburgers in less than ten minutes? Welcome to the world of competitive eating. Once a carnival spectacle, gluttony writ large is now an official sport administered by a global organization, the International Federation of Competitive Eating (IFOCE). The IFOCE was formed in 1997 and held twelve contests that year. Today it holds over one hundred events annually, pays out more than $40,000 in reward money, and has spawned an umbrella organization, Major League Eating, which sponsors worldwide events and handles merchandising and promotions. Smaller organizations such as the Association of Independent Competitive Eaters also hold contests throughout the year.

In competitive eating, contestants attempt to devour as many hot dogs, chicken wings, burritos, pancakes, or whatever the food du jour is, as fast as possible. Standard contest time limits are under fifteen minutes. The rules of the game strictly prohibit vomiting or, as it is referred to in the sport, "the Roman method" or "reversals." Any "reversal" during competition is an automatic disqualification. Common techniques for gorging as many hot dogs and buns (HDB), or other doughy items, as possible include "dunking," where the food is plunged into water or another liquid to soften it and make it easier to get it down, or breaking food like HDBs in half to get more into your mouth faster. "Chipmunking" is often seen in the last seconds of a competition, when contestants stuff their mouth with as many items as possible. Contestants are then usually given up to another two minutes to swallow. Bits of food that fall out of the eaters' mouths, called "debris," are expected to be "cleaned up" or the contestant loses points.

In 2002, the Fox network aired a two-hour special called *The Glutton Bowl* which featured three rounds of twelve-minute food trials. During the qualifying rounds, contestants had to eat as much mayonnaise, sticks of butter, hard-boiled eggs, hot dogs, hamburgers, and beef tongue as they could. In a wild card round, the semifinalists had to consume as many Rocky Mountain oysters, which were cooked but not fried, as possible, and in the final round they were challenged with cow brains. The winner, Takeru "Tsunami" Kobayashi, ate fifty-five brains in twelve minutes. In spite of being a gulletful of perverse entertainment, *The Glutton Bowl* has been held only once.

The most famous and long-standing celebration of gluttony is Nathan's Hot Dog Eating Contest. Held nearly every year since 1824 on Coney Island at the location of the original Nathan's Famous eatery, this July 4th competition pits approximately twenty contestants against one another in an all-out eating frenzy like no other. The ten-minute event has spawned eating stars like six-foot-tall, 218-pound Joseph Christian "Jaws" Chestnut, who in 2009 broke the world record when he gobbled sixty-eight HDBs in ten minutes. Chestnut is garlanded with many other extreme eating honors, including the world record for consuming 103 Krystal burgers in eight minutes.[33] Kobayashi (the *Glutton Bowl* champion) held the Nathan's hot dog scarfing world record from 2001 to 2006.

The world's best woman speed-eater is tiny Sonya "The Black Widow" Thomas. Born in Korea in 1967, Sonya has been an American competitive eater and female world champion since 2003. She set both the American and female world record for the Nathan's Hot Dog Eating Contest in 2005 when she downed thirty-seven HDBs in twelve minutes. Sonya earned her self-appointed moniker because she routinely annihilates men at least three times her body weight. As of 2010 she has broken and set her own world record in the women's category four times—most recently in 2009, when she wolfed forty-one HDBs in ten minutes. Sonya has also won numerous awards outside of the dog and bun category, including setting a world record for cheesecake consumption—eleven pounds in nine minutes. Most notably, on September 5, 2010, she won first place, beat Joey Chestnut, and set a new world record at the National Buffalo Wing Eating Contest in Buffalo, New York, when she ate 181 wings in twelve minutes.

Sonya is five feet five inches tall and weighs a mere 98 pounds soaking wet, and she wants to stay that way. Being very thin is actually

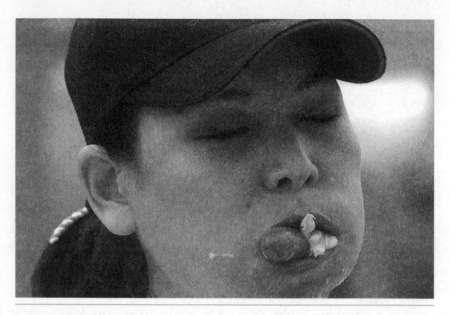

Figure 1.2

Sonya Thomas at the end of the Nathan's Hot Dog Eating Contest, July 4, 2007

an advantage for this sport, as it allows her stomach to swell without constriction from the ring of fat that surrounds the stomach of anyone with a paunch. To get in shape for her binges, she works out on a treadmill five days a week and eats only one big meal a day. For her day job, Sonya is a manager at Burger King, and so her typical repast includes several large orders of fries, a chicken Whopper, twenty chicken tenders, and two large diet soft drinks. When getting ready for a contest she includes a lot of fruit and vegetables in her meal and fasts the night before. Stomach elasticity is the most important asset for eating success, and the training methods of many contestants involve alternating fasting with stomach-stretching by drinking gallons of water, milk, and protein supplements and eating huge quantities of stomach-distending salads and vegetables.

As the nationality of the all-stars suggests, the sport of competitive eating has traditionally been most popular in Japan and the US. Recently, however, the sport has been viewed with disdain in Japan and it has also been criticized in the US for promoting gluttony during an obesity crisis, especially as it is attractive to teens, the demographic evincing the most alarming rise in obesity. There are also a host of health risks associated with the sport, especially damage to the stomach and digestive system. In addition to the possibility of internal bleeding, repeated excessive stretching of the stomach muscles can cause a condition called gastroparesis, where the stomach loses the ability to contract and empty itself. Nevertheless, the popularity of these uncouth displays of food exhibitionism illustrates how, like a moth to the flame, we are seduced by what we outwardly contend is revolting.

Food is a marvelous window through which to examine the multi-faceted emotion of disgust. First, food is a great passion, but it can also inspire terrible repulsion. Strangely, as with almost all facets of disgust, it is in our nature to be attracted to this repulsion. Who, uninitiated to the actual foodstuff, isn't just a little curious about tasting human flesh? Our attraction to what outwardly seems insane or revolting is a tendency called "benign masochism," which you will read more about later in this book.[34] Benign masochism is the uniquely human predilection for unpleasant or painful experiences when we can indulge in them safely. Eating habanero chilis until your face burns and your tongue goes numb with pain is a popular example.[35] Second, food disgust reveals that we are all disgustable, but what disgusts us specifically varies among us and it is all to do with what the object of our repulsion *means* to us. If lobsters are vermin, then lobsters are disgusting; if placenta is pâté, then placenta is tasty. Third, our food disgusts teach us that what is disgusting is subject to a wide variety of factors, from culture to context, that shape our interpretation of whether anything from chicha to chicken ice cream to tarantula pops is disgusting to you, or not. Finally, our disgust toward those who gorge on mayo, eat rotted shark meat, maggots, or human organ meat gives us a glimpse into one of the primary instigators of disgust—the recognition of our thinly veiled beastly nature. What else triggers disgust for us? The next chapter will give you a primer.

Chapter 2

A SNAPSHOT OF DISGUST

Marianne sat on the couch flipping through the pages of her boy-friend's "respectable" gentlemen's magazine. Glancing idly at the photos, the only feeling she had was a twinge of envy of the perfect bodies, until she turned to a page where, in the midst of the text, her eyes fell upon an image that made her body and mind twist in all-encompassing emotion. Marianne jerked her head back as far as she could and felt her face contort and her tongue escape her mouth. Could that porn star really be drinking urine out of a public toilet bowl? The picture caption proclaimed it to be an authentic act. The porn star was on her hands and knees, and she was collared and leashed. Her lips were pierced with dangling chains and her tongue hung salaciously into the yellowish liquid that she was lapping.

Charles Darwin first popularized the notion that there are discrete and universal human emotional experiences with corresponding facial expressions. In his book *The Expression of the Emotions in Man and Ani-*

mals (1872), Darwin wrote: "the young and the old of widely different races, both with man and animals, express the same state of mind by the same movements." The anthropologically oriented psychologist Paul Ekman, whose research fuels the popular TV show *Lie to Me*, continued Darwin's approach to understanding human emotional experience and, along with his colleague Wallace Friesen, traveled to Papua New Guinea in 1970 to test the idea that there were specific universal facial expressions of emotion. Ekman and Friesen presented members of the Fore Tribe—an isolated, modern Stone Age culture—with emotional stories and asked them to make facial expressions that corresponded to the emotional events. The Fore members readily did so, just as any literate Westerner would.[1] The legacy of this research is our knowledge that any healthy adult, regardless of their education, culture, or sophistication, can experience and recognize six universal emotions—anger, sadness, happiness, fear, disgust, and surprise—and that each of these emotions has a specific facial expression, physical response, and mental state associated with it. From Marianne's reaction, you no doubt know that she is disgusted.

Disgust grips our central nervous system and makes our blood pressure drop. It makes us sweat less and can trigger fainting, nausea, and vomiting. Outwardly, our limbs and core may shudder and recoil, and from our mouths the noises of "ewww" or "ughh" might escape. Disgust causes an array of mental states that vary from mild negativity to overwhelming revulsion, but they all center around the urge to pull away from, get rid of, and generally avoid that which is causing the feeling. Our faces also display a highly specified set of muscle contortions: the mouth opens (the tongue may also extend), the nose wrinkles, and the upper lip retracts while the lower lip protrudes. Even people who have been blind from birth make the same face of disgust as everyone else.[2]

One of the first and most interesting things I ever learned about disgust is that the face we make when we taste bitter melon is the same as the face you'd make if you were asked to hold your neighbor's freshly removed dentures. According to Paul Rozin, the father of the psycho-

Figure 2.1
The Face of Disgust

logical study of disgust and a professor at the University of Pennsylvania who has been studying the topic for over fifty years, our response to bitter taste is the sensory origin of the emotion of disgust, and all our other disgusts are built upon it.

The purpose of the bitter face is to promote either expulsion of what you've brought into your mouth or to keep anything that you might be contemplating putting inside you, out. This is why feeling disgust can make you stick out your tongue or even vomit. When we make the disgust face we also close our body, especially our face, off from invasion—which is why your upper lip retracts, your nose scrunches up, and your eyes squint. A scrunched-up nose decreases the amount of air that can flow into your nose, and squinting your eyes decreases your field of vision.[3] The disgust face helps promote one of the central functions of disgust—to keep the outside away from our inside.

It's a good thing that we don't usually follow through with swallowing when the bitter-taste face is upon us. Bitter taste signals that there are high levels of alkaloids in our mouth, and alkaloids tend to be poisonous, like lye and rotten foods.[4] Have you ever cracked open a walnut and been disturbed by its unexpected acrid taste? The walnut tastes bitter because it has started to spoil and become slightly toxic. Eating bitter nuts can give you a stomachache, but they won't kill you.

Disgust is a mystery relative to our other emotions. Experts in the field still do not agree on its fundamental purpose, precisely how it is processed in the brain, or even its universality. My aim in this book is to paint a picture of disgust as I have come to know it and allow you to reach your own conclusions. This chapter will lead you through the various types of disgust triggers that have been identified, from its basic origins in bitter taste and as a defense against eating toxic substances through to abstract levels of social repulsion at immoral acts, like cheating and incest. You will also find out how we develop the emotion of disgust, how our different senses inform us of what is disgusting, and how disgustable you personally are.

YOUR PERSONAL DISGUST

Our squeamishness quotient is a personality trait, just as how much of a party animal you are or how much you worry about the future are personality traits, and the specific entities that make us feel disgusted are also personal to each of us. What makes *you* feel like gagging or jerking away, or sends a shudder of revulsion through your core? Would it be the scent of raw, rotted shark meat? Meeting the Elephant Man? Shaking hands with a stranger who just coughed into his hands? Sitting on the bus next to a loudly belching stranger? The thought of your neighbor having sex with his daughter? Reading about the money mogul who defrauded his clients? Viewing the remains inside a freshly dug-up grave? Would it be all of these situations? Only a few? Or perhaps even none? Beyond pondering these questions, you can get a numerical measure of your personal predisposition to disgust by answering this questionnaire.

DISGUST SCALE[5]

Please indicate how much you agree with each of the following statements, or how true it is about you. Please write a number (0–4) to indicate your answer:

0 = Strongly disagree (very untrue about me)
1 = Mildly disagree (somewhat untrue about me)
2 = Neither agree nor disagree
3 = Mildly agree (somewhat true about me)
4 = Strongly agree (very true about me)

_____ 1. I might be willing to try eating monkey meat, under some circumstances.

_____ 2. It would bother me to be in a science class and see a human hand preserved in a jar.

_____ 3. It bothers me to hear someone clear a throat full of mucus.

_____ 4. I never let any part of my body touch the toilet seat in public restrooms.

_____ 5. I would go out of my way to avoid walking through a graveyard.

_____ 6. Seeing a cockroach in someone else's house doesn't bother me.

_____ 7. It would bother me tremendously to touch a dead body.

_____ 8. If I see someone vomit, it makes me sick to my stomach.

_____ 9. I probably would not go to my favorite restaurant if I found out that the cook had a cold.

_____ 10. It would not upset me at all to watch a person with a glass eye take the eye out of the socket.

_____ 11. It would bother me to see a rat run across my path in a park.

_____ 12. I would rather eat a piece of fruit than a piece of paper.

_____ 13. Even if I was hungry, I would not drink a bowl of my favorite soup if it had been stirred by a used but thoroughly washed flyswatter.

_____ 14. It would bother me to sleep in a nice hotel room if I knew that a man had died of a heart attack in that room the night before.

How disgusting would you find each of the following experiences? Please write a number (0–4) to indicate your answer:

0 = Not disgusting at all
1 = Slightly disgusting
2 = Moderately disgusting
3 = Very disgusting
4 = Extremely disgusting

_____ 15. You see maggots on a piece of meat in an outdoor garbage pail.

_____ 16. You see a person eating an apple with a knife and fork.

_____ 17. While you are walking through a tunnel under a railroad track, you smell urine.

_____ 18. You take a sip of soda, and then realize that you drank from the glass that an acquaintance of yours had been drinking from.

_____ 19. Your friend's pet cat dies, and you have to pick up the dead body with your bare hands.

_____ 20. You see someone put ketchup on vanilla ice cream, and eat it.

_____ 21. You see a man with his intestines exposed after an accident.

_____ 22. You discover that a friend of yours changes underwear only once a week.

_____ 23. A friend offers you a piece of chocolate shaped like dog doo.

_____ 24. You accidentally touch the ashes of a person who has been cremated.

_____ 25. You are about to drink a glass of milk when you smell that it is spoiled.

_____ 26. As part of a sex education class, you are required to inflate a new unlubricated condom, using your mouth.

_____ 27. You are walking barefoot on concrete, and you step on an earthworm.

WHAT'S YOUR DISGUST SENSITIVITY?

To calculate your disgust sensitivity score: First, put an X through your responses to items 12 and 16. These items don't count, they are just included to make sure you're paying attention. Then "reverse" your score on items 1, 6, and 10 by subtracting what you wrote from the number 4, and write those numbers in the margin. Finally, add up your responses to all twenty-five items (using your "reversed" scores on 1, 6, and 10). The total will be a number between 0 and 100. The average score on this test is about 40 (data from YourMorals.org).

This questionnaire, known as the Disgust Scale, was developed by the psychologist and author Jonathan Haidt (whose work you will encounter numerous times throughout this book), when he worked with Paul Rozin and Clark McCauley during his graduate school days at the University of Pennsylvania. The Disgust Scale is the gold standard test for measuring personal disgust sensitivity, and it has been used in hundreds of experiments and clinical tests since its development nearly two decades ago.[6] Whenever "disgust sensitivity" is mentioned in this book, it refers to scores on this test.[7] Your answers on this questionnaire will reveal to you how generally stoic or squeamish you are, as well as how much specific things such as moldy food, gore, germs, and death are repulsive to you.

How disgustable you are and what types of things are especially revolting to you is a fairly stable personality trait. If you answer the Disgust Scale today, your score will be close to your score on this test a year from now. Nevertheless, our specific disgusts do not stay the same within us across various situations and moods. For example, if between now and next year the media hypes some kind of infectious disease, like swine flu, your disgust response to germs could increase beyond your usual aversion to disease and contamination. By contrast, if you're feeling bold and carefree and in the company of hardy adventurists, you might decide to pick up the pizza slice that slipped off your plate onto the tavern floor and eat it, and to do so you'd have to dampen down any personal tendency to be disgusted by grime and contamination.

What specifically disgusts each of us also depends on how we "meet" the offending object. For example, you have saliva in your mouth now. Is it disgusting to you? Probably not. Now spit some into an empty glass and drink it. Doesn't that seem a little gross? How about a little saliva swapping with your lover in a passionate kiss? Pretty appealing, but oh . . . the unattractive waiter is talking to you and a splutter of his spittle lands on your cheek. How revolting! This is just one of hundreds of examples where the situation you're in and the condition of the trigger can change your reaction from delight to disgust and everything in between.

For better or worse, our disgust sensitivity wanes as we get older.[8] When Marianne is eighty-five, she'll be less bothered by anything she might come across in a "gentlemen's magazine" than she is at twenty-five. Just like our eyesight and our sense of smell, our sense of disgust diminishes as we enter our golden years. It is interesting to consider whether this may have anything to do with our diminishing ability to perceive the world around us. How our sensory capacities are related to our disgust sensitivities is not yet known, but it is the case that when we are eighty-five we've seen, heard, smelled, and generally experienced more of the disgusting things that life has to offer than when we are twenty-five. Exposure tends to lessen responses to disgust as well. For example, people who have had to care for someone sick and change their soiled bedding or dress their wounds are less disgusted by blood and bodily fluids than those of us who haven't had this background.[9]

Women are also generally more sensitive to disgust than men, which is why Marianne was more disgusted by the photo than her boyfriend was.[10] In nearly every study that has measured disgust sensitivity, women have been shown to be more strongly and more easily repulsed than their male counterparts. Intriguingly, in a recent brain imaging experiment women were, as expected, found to have greater brain activation than men to pictures showing people making disgusted faces. But men showed more activation to faces depicting contempt than women did.[11] Contempt is a social emotion that denotes hierarchy and superiority. If I look at you with contempt, I am signaling that I am better than you. Men are more sensitive to displays of social dominance than women are—even in a benign experiment where all that the subject does is look at a stranger's face, threats to hierarchy and dominance are more disturbing for men than they are for women.

We might at times feel that emotions are frivolous and unnecessary burdens—making us sad, tormenting us with anxiety, getting us into useless fights—and that we would be better served if we could put our feelings aside and get on with it. Philosophers such as René Descartes railed against emotions as the enemy of our intellectual capacities. But our emotions are critical for keeping us alive and procreating, which is the primary goal of our "selfish genes"—to pass along as much of our own genetic lineage into future generations as possible. Positive emotions, like happiness, keep us hale by propelling us toward happiness-evoking incentives; good food, sunshine, puppies, other happy people. Happiness makes us more engaged with the world and makes us want to approach what makes us feel good—such as people whom we sexually desire, thereby helping us to reproduce. Happiness also makes us more sexually desirable to others. If you feel good, you look better, and other people rate you as more physically attractive as well. For example, men who felt confident were rated by women who only saw them on videotape as being better looking than men of the same physical attractiveness who felt insecure.[12] Negative emotions keep us alive by protecting us from danger and death. Anger helps us to fight off a threat, fear and sadness help us to escape from or avoid threats, and surprise alerts us to something new and prepares us to take action. Disgust, first and foremost, makes us avoid; it compels us to withdraw from that which is disgusting.

Disgust, however, feels strange compared to our other emotions. It twists our body and our mind, and it can come and go to the very same entity by the whimsy of the situation or our cultural heritage. It can be stimulated by an entirely physical sensation, such as stepping barefoot into human excrement, and make our thoughts stop and our stomach heave, or it can be instigated by events that are purely conceptual, like corrupt politicians and Ponzi schemes, which feel intellectual and abstract. Moreover, it seems as though we could be disgusted by nearly anything, depending on how we think of it. If disgust has such varied triggers and qualities of experience, what is it?

The things that disgust us are not entirely universal, but they can be broken down into a set of general categories that most of us in the modern Western world can relate to. Each of these components serves as a facet of the emotion writ large, and Marianne's plight illustrates the various levels of disgust that define this emotion.

WHAT'S DISGUSTING?

The most basic form of disgust is about things that relate to our body. Body disgust is one step removed from our reaction to bitter taste. Body disgust has to do with what we might bring into our body, such as food and drink, and things that come out of our body: urine, vomit, phlegm, saliva, sweat, blood, pus, feces. The picture Marianne is looking at it is classically body-revolting because it combines bringing something into our mouth and a body fluid—drinking urine. The item on the Disgust

Scale that asks how true of you it is that "If I see someone vomit, it makes me sick to my stomach" taps into this type of disgust.

The one body fluid that seems exempt from our repulsion is tears. Although ancient Hindu scripture includes tears along with ear wax, eye mucus, and the usual suspects as part of a twelve-item list of body impurities, most of us don't think of tears as disgusting. This is because tears appear for all intents and purposes like water. Indeed, tears are 98.2 percent water. The remaining 1.8 percent is composed of proteins, oil, and excretory elements, but these body products are imperceptible to us. We equate water with purity, and therefore to the observer tears are inherently *clean*. Clean is considered the antithesis to disgust at several levels. Further, tears emerge from our eyes—the windows to our souls—and hence have a semi-spiritual property which is also connected to purity, not disgust. Tears are also typically emitted during sadness and stress, and we are usually compelled to comfort and console our distressed brethren. Consoling requires approaching. Children as young as two years of age are compelled to approach and soothe someone in distress, even if they don't know them.[13] If we were disgusted by tears, we would pull away and avoid our crying friends. It is therefore both to our personal advantage and good for social welfare in general that we are approachable when we are crying.

Body disgust is mostly triggered by our senses: putrid smells and tastes; slimy, oozy, red, brownish and yellowish sights; wriggly and wet touches; and squelching and retching noises. Stepping barefoot into human excrement is disgusting to all of our senses except, fortunately, taste. Our reactions to body disgusts are also body-centered, meaning that we feel disgusted by them primarily in our physical body—our gut in particular—rather than in our mind. In fact, body disgust often hijacks our mind so that we can barely think at all. But our mindless reactivity is sometimes in error, and we get that exasperated feeling of having been tricked by our emotional brain when we realize that the wet coiled thing on the ground was really only a shoelace and not an earthworm.

A closely related category of disgust concerns the nasties that we could "catch" from something or someone else, such as diseases or germs, as well as the supernatural foul essences that people or objects can emit. The unpleasantness of shaking the hand of a stranger who just coughed into it, noticing the scabby sores on the person sitting beside you, or being forced to work at a dirty, sticky desk are examples of disease–contamination disgust. The item from the Disgust Scale that asks how much you agree with the statement "I probably would not go to my favorite restaurant if I found out that the cook had a cold" taps into this type of disgust. The item on the questionnaire which asks about your willingness to "sleep in a hotel room where a man had died of a heart attack the night before" reflects contamination-type disgust in the supernatural realm. Disease–contamination disgust merges with body disgust, because a key reason for why bodily fluids are disgusting is because they may be disease-ridden. In the photograph Marianne was confronted with, the germs in urine and the other human waste the toilet bowl has contained are conjured in our mind and portend contamination and disease, which disgusts us. Long before germ theory was scientifically known, we were disgusted by other people's body fluids and wastes. We just didn't know how bathing or germs fit into the equation and in hindsight sometimes had some absurd ideas about the disease process. For example, during the height of the bubonic plague in Europe, people were instructed *not* to bathe because it was feared that cleansing would make the body softer and more vulnerable to infection.[14]

Another aspect of disgust that relates to the body is when the external human form is unusually distorted or mutilated. This includes physical deformities, gore, amputees, the very ugly and those who are morbidly obese or skeletally thin. The item on the Disgust Scale that asks you to rate the disgustingness of "seeing a man with his intestines exposed after an accident" taps into this type of disgust. Mutilation–deformity disgust is also elicited when normal bodies have been maimed in some way that we aren't attracted to, even when no blood or serious physical harm seems involved. Deliberate body piercings, scarification, and tattooing can be disgusting for these reasons. One current example that I personally find disgusting is the teenage trend of earlobe gauging. To achieve this look, your earlobe is first pierced with a 20- or 18-gauge needle—about the size of the tip of an average knitting needle—and then the hole is kept open with an equivalently sized earring. As a result of the skin's natural elasticity and with insertions of earrings of ever greater size, the earlobe gradually stretches and holes as large as a quarter or more can be achieved, though the fleshiness of your original earlobe is a limiting factor.

The model in the photograph that Marianne was looking at had her lips pierced and dangling with chains and so is a trigger for mutilation disgust. Mutilations, scarring, evisceration, blood, and gore are disgusting because they signal how tenuous and easily damageable and destructible our bodies are.

It also turns out that certain phobias—fear of critters like rats ("It would bother me to see a rat run across my path in a park), snakes and spiders, seeing needles inserted into you or someone else, and seeing or giving blood—involve fear and disgust in equal measure.[15] I have this reaction to having my blood taken. I absolutely cannot look as the needle is inserted into my vein because if I do I will become drenched in such a blend of dread and disgust that I might inadvertently flail and really hurt myself or the unsuspecting phlebotomist. I am not alone. About 10 percent of the population reports excessive distress at the thought of getting a needle injection or blood draw, and about 2 percent of the population meet the clinical criteria for having a bona fide phobia of these situations—meaning that their reaction negatively affects the way they go about their lives, impeding medical procedures or the decision to have children.[16]

In one recent study conducted at the Mayo Clinic, 3,300 patients aged nineteen to ninety-nine who came in to get their blood taken were given a questionnaire about their physical and psychological experience and history with blood draws.[17] From the answers it was found that those who reported being "disgusted" by needles were most likely to have a true phobia of getting their blood taken. By contrast, none of the more than 3,000 patients who didn't have a phobic reaction to blood draws were disgusted by needles. "Disgusted" needle phobics were also more likely to be younger and female, which is consistent with the findings that disgust is higher among women and declines as we get older.

Interestingly, in the Mayo Clinic study, the most important personal history factor that contributed to whether or not a person had a phobia of getting their blood taken was whether they had ever fainted or had near-fainting symptoms, like dizziness and nausea, while getting their blood taken in the past. If they had these symptoms, they were very likely to have a blood and injection phobia. The reason this is interesting is because dizziness, nausea, a drop in blood pressure, and fainting are the physiological correlates of disgust. This means that feeling the physical symptoms of disgust can make you feel emotional disgust at the stimuli that are associated with those physical symptoms: needles. This conditioned response of disgust toward needles then makes you feel light-headed and nauseous when you have close encounters with needles in the future (such as when you get your blood taken), because this is how your body feels when it is disgusted—and it becomes a vicious circle. Needles make you disgusted, disgust makes you feel dizzy, and a blood draw becomes ever more aversive and revolting. And, yes, I once nearly fainted while my blood was being taken, and it is only since then that I have been felled by the annual blood test.

Disgust and fear also ramp each other up. In a study that investigated the development of animal phobias, it was found that when school children in grades six through eight were given disgusting information about a fictitious Australian animal called a "quokka," such as that it "liked to grease its pelt in dung" and "eat maggots," they became more afraid of it as well as more disgusted by it. And when they were told how threatening the "quokka" was, for example that it had "very sharp nails with which he can hurt you badly," they became more disgusted by it as well as more frightened.[18] Not only does this show how feelings of disgust and fear can compound each other, forever amplifying and perpetuating a phobia, but also that the feelings of disgust and fear are fundamentally related.

The best way to rid yourself of these phobias is, unfortunately, to expose yourself a lot to your nemesis.[14] If you have a phobia of tarantulas, the best way to get over it is to get a tarantula for a pet. Clinically, exposure therapy is done in gradual steps of learning how to relax, confront, and accept your terrorizing-repulsive trigger. It ends with you stroking the beast's furry little back and the confident claim that you are no longer disgusted or frightened by these cute and indeed harmless spiders. If I wanted to cure myself of my blood and needle phobia, I should probably take a phlebotomy course or volunteer at my local Red Cross center.

In direct opposition to therapy which helps you overcome your repulsions, there is a psychological treatment in which disgust is deliberately used to encourage feeling horrible. Aptly named "aversion therapy," this is a technique where disgust or pain is paired with experiences that the person desires, but shouldn't. The idea is that if you make a bad behavior feel sickening or aversive, then you may stop the person from doing it again. Aversion therapy has been tried as a "cure" for homosexuality, and currently some prisons in the US are experimenting with aversive odors in an attempt to rehabilitate sex offenders.

Pamela Dalton, a scent psychology expert from the Monell Chemical Senses Center in Philadelphia, has been working with some correctional institutions to assess the effectiveness of odor aversion therapy. A pedophile, for example, is presented with pictures and sounds that are specifically chosen to be erotic to him and at the same time exposed to a nauseating or painful odor. In the past ammonia was frequently used because it is both painful and unpleasant-smelling, but Pam Dalton is working with novel odor combinations that are intended to be especially disgusting. The goal is that whenever the pedophile then encounters a "wrong" sexual trigger, the aversive association from the odor will overwhelm him with disgust and he won't be able to proceed. When the treatment sessions are over, the offender is typically given a vial of the nauseating odor to take with him, with instructions to sniff it whenever he feels aroused by things he shouldn't be. Another supplement to the treatment is to pair pictures and sounds of appropriate sexual interactions with a clean, floral or minty scent, so that the offender learns to associate appropriate sex with pure and positive emotions. Evaluation of this technique is presently ongoing so it is not yet known if, or how well, odor aversion therapy works.[20]

THINKING MAKES IT SO

The types of disgust triggers I've described so far all have to do with noxious encounters at a direct physical level. These triggers don't usually require much thought or analysis, but as we move up the ladder of disgust we come to a set of elicitors where we do have to engage our mental faculties relatively intensely. These elicitors are disgusting because of what they *mean* to us. As mentioned in the last chapter, one of the most meaningful triggers for disgust is any experience or entity that reminds us that—like grasshoppers, gophers, chihuahuas, and orangutans—we are animals. This is repelling because, like gore and deformity, it reminds us that we have a fleeting, finite, and fragile existence. This is also why we are disgusted by dead things ("I would go out of my way to avoid walking through a graveyard.") Death unequivocally confronts us with our animal destiny. But animal-reminder disgust is also more subtle.

Our desire to think of ourselves as unique among the rest of the world's creatures that have a beating heart fuels our repugnance for behaviors that are animal-like—behaviors that break the codes of manners and norms of "civilized" public behavior. To maintain our separation from lowly beasts, we should not be seen, especially in public, behaving like beasts: gobbling down sticks of butter, having sex, grunting, farting, belching, being naked, scratching ourselves, picking our toes, being unwashed, and squatting to urinate or defecate wherever the urge compels us. We are disgusted by people who behave like this, because they do not uphold the conventions of civility that we expect and respect, and because by breaking these rules they bring our inherent animal nature to the fore.

Belching, urinating wherever one pleases, and so on may seem obviously physically disgusting, but depending on the mores of the time or place these acts are or have been entirely acceptable in polite society. In medieval palaces, urinating and defecating against the dining room wall during dinner was the norm. In fact, in the palaces of yore, rather than having the servants clean up the mess, after it had piled too high people just moved.[21] Today, belching loudly is a gracious compliment to the host at the end of a meal in the Middle East. And though gobbling and ripping flesh from bone with bare hands and teeth may sound like the vulgar antics of some cartoon Viking, many of us frequently eat like this. Do you neatly cut up your chicken wings with a knife and fork and then carefully place one small morsel at a time into your mouth? Or do you grab a chicken wing with your fingers, gnash and suck the flesh and skin from bone and when you're done toss the remains, licking your fingers as you go?

Besides being revolted by anyone who acts like an animal, we are repulsed at the thought of being treated like animals and especially appearing to enjoy it. In the magazine photo that crushed Marianne, the model is posed on all fours, naked, collared, and leashed—just like an animal. Moreover, her salacious expression denotes that she is accepting, pleased, and possibly turned on by her situation. This is disgusting because it reminds us that not only are we made of flesh and blood like animals and have the same bodily functions and urges as they have, but that we can embrace our animality and *be* animals and therefore truly are no different from beasts on all fours.

Closely connected to our repulsion at animality is our disgust at sexuality. The item on the questionnaire that asks how disgusted you'd be if "as part of a sex education class, you are required to inflate a new unlubricated condom, using your mouth" taps into sexual disgust as well as basic body disgust. Sex is among the most brutish behaviors we engage in, and though all cultures have rules for "proper" sex, we cannot escape the fact that sex itself is raw and physical and that we make the beast with two backs quite literally.

Besides being more generally squeamish than men, women tend to be specifically more disgusted by sex than men are. For example, in the first experiment to investigate differences between the sexes in response to sexually explicit films, female college students rated pornographic movies as 20 percent more disgusting and less enjoyable than men did.[22] Many subsequent studies have corroborated the finding that women are more disgusted by sex than men are.[23] Marianne's repugnance for the photo is therefore implicitly tinged with a general female aversion toward overt displays of sexuality.

In addition to being linked to our animal repulsions, our sexual disgusts are highly entwined with our morals. Moral justifications are then used to bolster the rules for what sexual behaviors are acceptable or disgusting. Aside from heterosexuals having sex beneath the sheets, in the missionary position, for procreation only, our attitudes to sexual behavior—whether it be masturbation, a penchant for group sex, sado-masochism, pornography, homosexuality, incest, or necrophilia—are colored by our beliefs about what kind of sex is "right" and what is "wrong."

This brings me to the most complicated type of trigger for disgust: our disgust at moral violations. We say "that's disgusting" when people behave in ways we consider animalistic and sexually debauched, but also when we hear about the rich politician who pilfered money intended for an after-school program to renovate his mansion. Moral disgust requires a very sophisticated level of thinking. Indeed, fierce arguments rage among mature and intelligent adults about what is morally disgusting or not. Moral disgust forms the basis of ideologies from politics to religion. Likewise, moral disgust is influenced by our religious and political convictions. Marianne might be more or less disgusted by the image of the porn star depending on her religious adherence and whether she is politically conservative or liberal. The more regularly she attends church and the more conservative she is, the more repulsed she'll be by the photo.[24] Similarly, moral disgust is heavily influenced by culture. What is deemed appropriate by one group can be perceived as disgusting by another—whether it be using your left hand to take a meatball from the communal dinner plate to stoning an alleged adulterer to death. It is also possible that moral disgust may not be disgust at all. Rather, as you will discover in chapter 8, it may be that when we say we are disgusted by sleazy politicians we are really feeling another emotion altogether.

Many triggers for disgust are not simple hits, but rather intersect and activate multiple layers and levels of disgust. The picture Marianne was so repulsed by activated all of them—body disgust, disease–contamination disgust, mutilation–deformity disgust, animal, sexual, and moral disgust. Our disgusts are also so pervasive and often feel so basic, obvious and instinctual—who wouldn't be disgusted by the smell of spoiled milk, as the Disgust Scale asks—that it begs the question: are we born with the sense of disgust?

THE INSTINCT THAT HAS TO BE LEARNED

> One day when Kathy was four years old, and strolling her doll carriage down the street near her house she found a "poor, sick kitty." Kathy picked up the sick kitty, put it in her carriage, and took it home for her Mommy to make it better. But when Kathy's mother saw the stiff, dead skunk swaddled in the doll pram she shrieked to high heaven and dragged Kathy into the bathroom for a vigorous scrub-down. The pram was never seen again.

Disgust is the last emotion children acquire. Happiness comes first, followed by sadness or distress, and both of these emotions appear very early on, as any parent knows. By four months of age children can get angry, and by the time they're six to seven months old they can feel fear and surprise. But despite the fact that your bundle of joy will grimace and spit out things that taste bitter from birth, she won't experience any other form of disgust until she is at least three years old. And it's not until she's at least five that she'll be able to have an inkling that the look on your face means that you're disgusted by the dead skunk she's just presented you with. In fact, she's most likely to think that you're angry with her—which in the case of Kathy's skunk story might also be true.

But even in cases of pure yuckiness, such as seeing someone eat a wormy apple, a recent experiment showed that children as old as nine (the oldest tested) were only able to recognize the facial expression of disgust, made by either a child or an adult as being "disgust" 30 percent of the time, and only explained the face as being due to something yucky 14 percent of time. Instead, they were most likely to say the expression on the face and the cause of the face was "anger." Indeed, they were just as likely to interpret the disgust face as being an angry face as they were to interpret a true angry face as angry.[25] In fact, the faces of anger and disgust are most likely to be confused for each other throughout our lives,[26] which, as you'll discover toward the end of this book, has some intriguing implications for understanding what "moral disgust" really might be.

Our early oblivion to disgust may be adaptive because mucking around in what's dirty, such as dirt, strengthens our immune systems. Dirt is teeming with bacteria. In fact, most medical antibiotics are made from bacteria that come from dirt. Just as getting inoculated with a vaccine enables our body to mount an immune response so that we can fight off the real thing, exposure to benign bacteria stimulates our immune system so that it is better able to fight back when bad bacteria are encountered. The old-fashioned approach of letting toddlers crawl around on the floor and outdoors actually results in better overall health when children are older, including lower rates of asthma and even type 1 diabetes, than the current fashion of hyper-hygienic child care.[27]

We have to learn what is disgusting, but we don't *get it* all at once. Rather, our ability to learn and feel various forms of disgust coincides with the ability of our brains to process and comprehend the various degrees of complexity in which disgust manifests. The first disgust lesson most children are taught, and why disgust emerges around age three, is toilet training. Prior to age three, many children will finger-paint with and even eat their own excrement. After toilet training and as our mental capacities increase, our ability to understand and see disgust in various situations emerges, and intriguingly your youngster's ability to discriminate between your angry face and your disgust face coincides with his ability to say "that's disgusting."[28] But it isn't until mid- to late childhood that we can understand why our animality and death should disgust us, and not until adolescence or possibly beyond that we feel disgusted by a rich city official stealing money intended for an after-school program. Our understanding and experience of disgust tracks our intellectual development.

More than any other emotion, children must be exposed to and trained in the specific cultural and social norms of disgust, whatever they may be. Differences in culture manifest in different attitudes toward what is disgusting. In North America, toilet training is taken very seriously, and poop is treated as the worst of all bodily abominations. Despite its daily necessity, the message is that defecating is to be done as imperceptibly as possible—the main reason bathroom air fresheners exist is to eliminate any olfactory evidence of this event. Indeed, we condemn poop with such repugnance that many Western children, once they have been potty trained, will show excessive disgust toward toilet activities. William Miller, author of *The Anatomy of Disgust*, observed this in his own children and described how his young daughter, after being toilet trained, refused to wipe her behind for fear of contaminating her hand.[29] I personally know someone who during adolescence developed such severe constipation that she had to be hospitalized because for years she had been refusing, as much as physically possible, to have any bowel movements because she found them so disgusting. Our disgust toward things like slugs, snails, and other moist, coiled objects occurs because they look like feces.

In several non-Western cultures, toilet activities are viewed with more acceptance. For example, an anthropologist told me that in some tribal villages in Cameroon squatting down with the proverbial newspaper is customarily done in outdoor public settings. However, other bodily fluids are seen as more revolting than westerners typically consider them to be, in particular menstrual blood, which you'll find out more about in chapter 7.[30]

Another potent illustration of how disgust has to be learned comes from the observation that wild or "feral" children who have grown up without any human contact or socialization and obviously no toilet training, don't appear to show disgust when confronted with the things and situations that inspire revulsion in the rest of us.[31] The most famous case of a feral child, the Wild Boy of Aveyron, who greatly disgusted the doctor who studied him, apparently had no awareness of disgust beyond the basic reaction of avoiding rotted meat: "A dead canary was given him, and in an instant he stripped off its feathers, great and small, tore it open with his nails, smelt it, and threw it away."[32]

Not only is disgust not innate, we are also the only creature that experiences it. Your new puppy will love you just as fiercely whether you look like George Clooney or are grotesquely disfigured. Many dog owners will be familiar with their pet's unfortunate delight at rolling around in, or playing with, a multitude of foul-smelling excrement-based or dead toys. Even chimps and gorillas that appear to share all our other basic emotions do not exhibit facial expressions or behaviors that anthropologists equate to disgust in humans.[33] The only element of disgust that dogs and primates show is the reaction of distaste (the etymological meaning of the word disgust is "bad taste," from the French *dégoût*, originally from Latin *dis* = negative, *goust* = taste). Like infants and wild children, animals will react to foods that taste bitter and steer clear of foods that have made them sick in the past, but they do not experience emotional disgust.

FASCINATED BY DISGUST

As children acquire an understanding of disgust, their responses change from hypervigilance to fascination, and around puberty, a giddy enthrallment with "disgusting" things is raging. This is why disgust has been used so successfully for both commercial and educational benefit with this age group. The toy company Fundex has capitalized on the understanding that kids love grossness. What's In Ned's Head is, according to Fundex, "the gross-out game kids love so much that they voted it #1!" Ned's head contains fake vomit, snot, moldy cheese, and a rat, among other queasy items, and the hands-on game makes kids reach inside Ned's foam nostrils, mouth, and ears to find these delights.

Then there's the aptly named board game Totally Gross, developed by University Games Corporation. The game brims with disgust and includes instructions like "land on a Gross-Out space, and you may have to check another player for toe jam or describe the last time you threw up!" Science education is the pitch for this game, whose makers proclaim, "A dose of gross helps science make sense!" Education was also the theme behind the very successful Canadian animated television series *Grossology*, loosely based on the book with the same name (the show aired in the US on Discovery Kids Channel). The plot involves a brother and sister crime-fighting team who work for a secret government organization called the Department of Grossology, and battle cleverly named villains such as the Scab Fairy. The duo's adventures impart scientific lessons and facts; in one buggy episode, it was revealed that termites communicate through scent.

Repulsiveness can even get kids excited about art.[34] An exhibition showcasing "maggot art" was held to great acclaim at the USA Science and Engineering Festival on the National Mall in October 2010. To create maggot art, diluted water-soluble paint is dropped onto a canvas and maggots are then deposited onto the paint drops. As the maggots wiggle around they leave Jackson Pollock-like designs in their wake. The creator, Erin Watson, a biologist from the University of Southeastern Louisiana, teaches children about maggots and their importance in decomposition when she gives her artistic demonstrations. Kids apparently can't wait to get involved in the art-making process with her.

Any parent who notices what their children are doing or saying as they mature through their tween years can recount dozens of nauseating stories involving bodily secretions. One of the interesting facts about these disgusting delights is that most of the bodily fluids that children use to taunt others with are not disgusting to the one who is doing the taunting. "Here's my snot and I'm going to try to smear it on you . . ." The stuff from me is fine, but the stuff from you isn't. This "I'm okay, you're not okay" reaction illustrates one of the fundamental functions of disgust—to protect the boundaries of our physical self from outside invasion.[35] We know that we can't contaminate ourselves with what is inside us; it's already there. This is why the saliva in your mouth isn't offensive. But it can become contaminated by outside "other" yuckiness as soon as it leaves your body, which is why imagining taking a drink of your own spit can make you wince. It is far worse, however, if the spit comes from someone else; this is why drinking chicha seems abominable. Though adults are not nearly as cavalier as children, the "I'm okay, you're not okay" reaction persists into adulthood; we still admire our own bowel movements, but gag if we see another's. This may seem like one of those experiments that really didn't need to be conducted, but Richard Stevenson, a smell and disgust expert at Macquarie University in Australia, and his colleague Betty Repacholi of the University of Washington in Seattle collected data from unabashed volunteers and scientifically documented that we have more negative reactions to the smell of other people's poop, farts, foot odor, and sweat than our own.[36] Richard Stevenson and his colleagues also showed that moms will declare that their own infant's poop smells fine but a soiled diaper from another infant of the same age is a total stench.[37] We even prefer the smell of our dog's poop to that of other dogs.[38] Our babies and our dogs are under our control and in our "ownership" and therefore feel like extensions of our selves.

As adults, though we may tolerate or even privately like the waste effluvia that comes from our own bodies or those whom we consider to be part of us—we don't go about inflicting our personal pleasures on others. So why are children so perversely beguiled by torturing their peers with disgust—body fluid based and otherwise? Is it for humor? Is it to purge the feelings of disgust from themselves? Is it simply to torment? Did you ever do it? I did.

When I was about ten, I discovered an unparalleled horror and then obsession, a book on my parents' bookshelf called *Animals Without Backbones*.[39] One afternoon, sitting on the couch idly flipping the pages of the black and white tome, my fingers flicked to a page that made me throw the book into the air and myself off the couch in a body-shuddering convulsion. What I saw was a picture of a man with elephantiasis, a limb-deforming disease caused by a parasitic filarial roundworm. In the picture, the man had one enormous and grotesquely deformed leg connected to an otherwise wraith-thin body. But when I had gathered myself and the book from the floor, I of all things, looked for the picture *again*, and I did this repeatedly, almost obsessively, for at least a year. My method was to flip the pages so that all I could see was a tiny bit of the bottom corner—and when I thought I had the page with the frightful image I would slowly peel it back to see if I was right, and I did this until I found it. My goal was then to keep looking at the picture for as long as I could stand it. I didn't know it at the time, but I was using exposure therapy to cure myself of my disgust phobia of this image. When I finally got to the point of being able to confront the page triumphantly, I began to sadistically make my friends look at the picture. And I remember feeling gleeful when they turned their heads away squealing, "Gross!" and I could stand there solidly staring at it.

Disgust is a great attention-getter, which for children is a key motivator of behavior. Playing with disgust also enables children to test the responses they get for that attention and where boundaries are drawn—when does my showing you a revolting picture or sticking my snot on you cease being playful or amusing and get you angry? Disgust is also a safe way to wage tit-for-tat antics. I am much less likely to really get hurt, unless you have a lethal contagious disease, if I smear some body fluid on you and you do it back to me compared to if I hit you with a baseball bat and then you hit me back. Still, there has been no direct research to figure out why the enticement of disgust lures kids to learn science, unleash their inner artist, promote the toy and game industry, and torture their friends, but I have a theory.

When we are young, unedited and unfettered by societal mores, we experiment with the limits of social behavior, the things that threaten us, the visceral, sensory, and sensual world all around us. Disgust is at the heart of these fascinations. Though we learn to turn off our outward zeal for these fascinations, the questions, temptations, and fears never go away. This is why we remain lured by disgust throughout our lives.

As children, part of this curiosity is about discovering our bodies, another is about testing the rules of the culture we live in and what is "right" and "wrong" to do, and a further dimension has to do with teasing ourselves with the fine line between disgust and desire and the as yet undefined boundaries between life and death. A child's wonder, curiosity, and pleasure in disgust may hold up a mirror to the private, uncivilized, animalistic part of our adult souls. Indeed, most adults engage in "gross" behavior all the time.

In 2005, a poll taken by the Response Insurance company found that 17 percent of Americans admitted to having nearly caused an automobile accident because of picking their nose while driving.[40] My guess is that the number of drivers who have safely nose-mined at some point in time is near 100 percent. Nonetheless, nose-picking while behind the wheel has been deemed to be more dangerous than using your cell phone, and in Canada a new law could impose an $850 fine for a moving nose-picking violation—$350 more than the fine for a cell phone driving violation.[41]

What if we could shed the veneer of niceties and civilized behavior that binds us to our social world? Would we revel freely in our body fluids or those of others? Some people do. Rapunzel syndrome is a disorder where people are compulsively drawn to eating hair—anybody's hair. They will go into your bathroom and suck out your hairbrushes. The desire to eat hair can become so severe as to require stomach surgery to remove giant obstructive hairballs. This condition takes yielding to base physical cravings a few steps too far. But Freud recognized that the human paradox of animalistic drives battling surface control was central to human nature, that it was in fact the cause of most of the neuroses that his adult patients suffered from. Children don't have this problem. The burden of maturity is having to put away our toys, whether they be manufactured or bodily produced.

SENSING DISGUST

The "I'm okay, you're not okay" reaction to body products and fluids is an example of how the source of the offending entity determines whether we think it is disgusting or not, and that the more something is a part of *you* the less disgusting it is. But when it comes to feeling that somebody else's feces are disgusting, what aspect of our ability to sense the world around us gives us the biggest hit? Is it the smell of it? The sight of it? The thought of touching it? How do our senses inform us and move us toward disgust, and which sense does it to us the worst?

It might seem that smells are most disgust-inducing because they are so enveloping. You can close your eyes, plug your ears, spit something out, or move your hand to get away from other sensory disgusts, but with a smell it is all around you and you have to move your whole body a fair distance to escape. Yet sniff information can be misinformation.

Our reactions to smells are not innate. Like the emotion of disgust, the joy or grief we experience from odors is learned and sensitive to complex influences.[42] The smell of spoiled milk isn't a "rotten" smell until you learn what good milk smells like and what the smell cue is for when milk has gone off. When you know that the milk has spoiled, the smell that coincides with that fact becomes a bad smell. It may seem hard to believe that feces don't inherently smell disgusting and lilies heavenly. But the scent of feces is only revolting once you've learned that feces means waste, and it varies in pleasantness depending upon whose you think it is. Personal history and first experiences also powerfully determine the meaning of odors. My mother hates the scent of lilies because her original associations to them were hospitals and death, rather than boyfriends and bouquets. The context in which we encounter an odor is a further influence. If a certain aromatic concoction hits your nostrils and you are in a Michelin three-star restaurant and see a waiter wielding a cheese cart your mouth might water, but if you were walking in the back alley behind a dive bar and the same aroma wafted by you might want to throw up. Even words can alter our perception of smells. My laboratory showed that just by calling a chemical mixture either "vomit" or "parmesan cheese" we not only could elicit totally different reactions to the scent—disgust or pleasure—but people wouldn't believe that they were actually smelling the same odor.[43] Smell is ambiguous and we can be misled to be disgusted by our noses.

Sound, like smell, is also indefinite and we need additional information to determine whether we should be disgusted or enchanted. We also have to learn the meaning of most noises in order to know what they signify and then decide whether or not to be repulsed. Squelching sounds aren't disgusting until we know that it means someone has stepped in a foul substance. The problem is that many noises have multiple significers. The sound of pitter-patter from above could be rain on the roof or rats in the attic. A squeak may be the scuff of someone's shoe or their fart. Without additional information about the noise in question, we can be deceived. Touch is similarly uncertain. I am blindfolded and my hand is being inserted into a bowl. I feel something round, wet, and with a firm but yielding density. Am I touching eyeballs or peeled grapes? Whether I am at a Halloween party or being tortured will determine my reaction.

By contrast, taste is definitive and does not require additional sensory cues or outside information to trigger a "correct" response. Bitter inherently tastes bad and we are compelled to spit it out. This reaction is instinctive, and we continue to be repelled by bitter even when we know it is nutritious. If you are very sensitive to bitter, you can be well aware that bitter greens are healthy and yet still refuse to eat them. Sight is not instinctive, but it is definitive. Once we have learned what rotted meat and dead skunks look like and what their significance is, there is no question that they are revolting. If you open a container from the refrigerator and see chicken legs covered in furry black mold, you have no doubt that the meat is rotted and you don't need to check any of your other senses to be sure. What you see is what you get.

Our eyes tell us whether the tickle across our toes is wriggling centipedes or the cat's paw, whether an aroma is coming from a coveted cheese or vomit, and that we are listening to rain not little rat feet.

Figure 2.2

Hieronymus Bosch, *The Garden of Earthly Delights*, **c. 1500. Detail from lower right corner, center panel**

Visual disgust is even stronger than the unpleasantness of licking a bar of soap, because though bitter tastes awful, taste is not usually emotional enveloping but rather only a sensory experience that tends to stay on our tongues. Tastes don't invade our mind the way sights do. By contrast, vision stirs our emotions and can elicit overwhelming and absolute disgust. It was only a photograph that devastated Marianne, and me, with repulsion. We all experience exquisitely disgusting nonvisual sensations, but vision gives us our worst disgust because we get the clearest meaning from what we see and because we rely and believe our eyes above all our other senses.

There is not one disgust, but many disgusts that vary from simple and physical to abstract and complex. The emotion of disgust is probably universal but it is not innate; disgust has to be learned and is subject to a myriad of influences. Our age, our personality, our culture, our thoughts and beliefs, our mood, our morals, whom we're with, where we are, and which of our senses is giving us the feeling, all shape whether and how strongly we are able to feel disgusted, both as a personal predisposition and in the moment right now. Disgust is also a window into our lurid enticements. Anyone who is intrigued by Hieronymus Bosch's famous painting *The Garden of Earthly Delights* understands how we are drawn to what is unseemly and must fight to control our inner animal passions. In sum, disgust is uniquely personal, highly psychological, culturally malleable, and contextually capricious. Nevertheless, as you will discover in the next chapter, disgust is dependent upon specific neural hardware. In order to feel disgust, no matter who we are or where we live, we must have an intact and healthy brain.

Chapter 3

DISGUST ON THE BRAIN

If you grew up in North America, there is a good chance that your child-hood memories include school bus trips with singalongs to "This Land Is Your Land"—one of the most famous folk songs in American history. Its lyrics were written by Woody Guthrie in 1940 and its unforgettable melody is based on the old gospel hymn "Oh, My Loving Brother." The song is so catchy and inspiring that when I was growing up north of the border in Montreal we sang a slightly modified version to suit Canadian geography. Besides childhood bus trips, you may also know hipper ver-sions of the song performed by Bob Dylan, Peter Paul and Mary, Bruce Springsteen, Dave Matthews, and the Counting Crows.

Woodrow "Woody" Wilson Guthrie was born on July 14, 1912, in Okemah, Oklahoma. Guthrie was married three times and fathered eight children, including the folk musician Arlo Guthrie who starred in the film based on Woody's song "Alice's Restaurant." Woody Guthrie inspired a generation of folk singers and would have inspired more, but at the age of fifty-five, on October 3, 1967, he died from complications of a rare and devastating disease which he had been suffering from since his late thirties, Huntington's chorea.

Huntington's chorea, also known as Huntington's disease (HD), is a genetic disorder that produces wasting and degeneration of nerve cells in an area of the brain called the basal ganglia—a set of interconnected neuron bundles that are critically involved in muscle and motor movement and coordination. Because of the breakdown in the brain areas responsible for movement, HD is most notably characterized by bizarre physical behavior. The symptoms initially come on as a general lack of coordination, an unsteady gait, and slurred speech. As the disease worsens, dance-like (chorea) uncontrollable and irregular muscle movements, especially of the arms, legs, and face, appear. Not surprisingly, people afflicted with this disease were frequently persecuted for witchcraft and demonic possession. Guthrie himself received diagnoses ranging from alcoholism to schizophrenia until his illness was finally recognized.

We owe our knowledge of this disease to George Huntington. In 1630, the Huntington family settled in East Hampton, Long Island— one of a series of moves they were forced on throughout the northeastern states. Their repeated resettlements were due to persistent accusations of devil possession and witchcraft because many children in the family seemed to be cursed by a strange and frightening demeanor. George Huntington, a third-generation New York descendant, had just earned his medical degree and was preparing to join his father's medical practice in Long Island when it struck him that the afflictions of his relatives were connected. The genetic mechanisms of disease weren't understood at the time, but Huntington recognized that the peculiar symptoms of his relatives were consistent and therefore must have been inherited. A year after leaving medical school, in 1872, he presented his observations on his family's disorder at a medical society meeting in Middleport, Ohio. His paper "On Chorea" was published in the *Medical and Surgical Reporter: A Weekly Journal* and remains the gold-standard description of this illness today.[1]

George Huntington's conclusions were correct, as we now know that the disease is inherited—but it is much more pernicious than typical genetic traits. The gene for HD is dominant, meaning that unlike most genetic malice, where both parents have to contribute a copy of their defective gene for their child to suffer the force of the mutation, with HD all you need is one parent to have the gene and there is a 50 percent chance that their child will have it too. And if you have the gene, you have the disease. The illness is progressive with no remission and is always fatal. As the disease advances, motor function gets progressively worse and every action that requires muscle control is in line to disintegrate, including speaking, chewing, and swallowing. Mental confusion usually sets in during the later phase of the illness, and without a constant caretaker, an HD sufferer won't live long once the symptoms become severe. Not surprisingly, suicide is a common cause of early demise among HD patients.

The symptoms of HD usually develop between the ages of thirty-five and fifty, and since this is often after the birth of children, many people transmit their gene for it without knowing. In addition to passing on his talent and love of music, Woody handed down his HD gene to two of his daughters, Gwendolyn and Sue, both of whom died when they were forty-one.

Because of lack of awareness and information about the disease at the time, Woody himself was essentially untreated. One positive outcome, however, is that because of the dedication of his wife, Marjorie, and Woody's fame, public consciousness grew and the Huntington's Disease Society of America was founded. Although there is still no cure, there are treatments which make coping with the disease better, as well as genetic testing to help those at risk manage transmission to future generations.

Curiously, in addition to the movement aberrations of HD, the illness is characterized by a strange and specific emotional deficit. Woody wouldn't have known what your reaction would be to seeing lips pierced with chains or drinking somebody else's fermented saliva. More importantly, he wouldn't understand the face you're making right now. This is because people with HD can't recognize disgust. What follows is the serendipitous and surprising story of how we have come to know this.

THE DISGUST-FREE DISEASE

Reiner Sprengelmeyer began his research career in the early 1990s at the German Science Foundation studying decision-making and attention in patients with Alzheimer's and Parkinson's diseases. In 1994 he received a one-year fellowship to travel to Cambridge, England, and examine patients with other brain disorders that affected cognition. It was here that he met and got excited by the work of Andy Young, a renowned neuropsychologist. Young's specialty was the neural circuitry involved in our ability to recognize human faces and facial expressions of emotion. At the same time in America, Ralph Adolphs, a budding star in the neurobiology of emotion, had just reported an unusual finding from a patient suffering from Urbach–Wiethe disease.

Urbach–Wiethe is a very rare hereditary disorder that causes waxy thickening of the skin, mouth, and throat. People with this disorder have problems speaking, can't cry, and often fly into fits of rage. The part of the brain that is most damaged by Urbach–Wiethe is the amygdala, an almond-shaped structure in the limbic system—the ancient core of our brain. The patient in question, a thirty-year-old woman dubbed SM, was of normal intelligence and cheerful disposition. Adolphs and his colleagues had been testing SM on a number of cognitive tasks, including several face and emotional expression recognition tests, and she generally performed well, but to Adolphs's surprise she had a peculiar problem. SM couldn't recognize faces displaying fear, though her ability to recognize other emotional expressions was nearly normal.[2] The reason this finding was so important is because it implied that specific brain regions controlled specific emotions. The amygdala, it seemed, governed fear.[3]

When Sprengelmeyer heard about Adolphs's discovery, his research took a turn. Sprengelmeyer knew that Huntington's disease caused damage to the amygdala as well, so he and Andy Young teamed up to test whether HD patients might also be unable to recognize fearful faces. Together with several colleagues, they compared the performance of a group of HD patients with a group of healthy adults matched for age, education, and intelligence. In this landmark experiment, HD patients and healthy adults were presented with various headshots and asked to figure out the person's age category (young or old), gender (male or female), familiarity (the face matches another face in a larger set), and the direction the person was looking (frontal or not). On all of these tests, the HD patients were normal and generally no different from the healthy adults. Next, the two groups were shown pictures of people making the six basic expressions of emotion (happiness, sadness, anger, fear, disgust, and surprise), and asked to indicate what each facial expression was. Here, unlike in the other tests, the HD patients were oddly different from the others. They were worse at recognizing all facial expressions except happiness, but there was one emotion that they couldn't see at all. It wasn't fear—it was disgust.[4]

In Sprengelmeyer's study, each facial expression was shown ten times, so if a participant recognized the emotion in the expression correctly each time, their score would be 10. The HD patients' average score for recognizing happiness was 9.5, nearly the same as the 9.9 average score for the healthy adults. However, when they looked at faces showing disgust, the HD patients scored an average of 1.9 compared to 8.1 for the healthy adults. None of the HD patients' score for disgust were better than random guessing, and six of the eleven HD patients either never got it right or picked "disgust" only once in ten times. Sprengelmeyer and his colleagues were stunned by his finding. No one had ever shown that people with a particular infirmity couldn't see disgust. Today, this curious deficit among HD patients has been replicated many times.

Recent studies have further found that besides not being able to recognize disgusted faces, HD patients are bad at identifying what the sounds of retching indicate, they're bad at identifying what emotion the sight of cockroaches or mutilation should induce, and they are less offended by the smell of feces than healthy volunteers.[5] In contrast to these oddities with disgust, the same HD patients are normal for detecting and responding to the sounds and sights of happiness, sadness, fear, anger, and surprise. Even more astonishing is that well before ever showing any physical symptoms of HD, carriers of the HD gene can't recognize disgust either.

In a study by Sprengelmeyer and his colleagues at the University of St. Andrews in Scotland, young adult carriers of HD who had not yet developed any symptoms were compared for their ability to recognize the six classic faces of emotion against peers of the same age with one HD parent but who had themselves been spared the HD gene.[6] Like patients with full-blown HD symptoms, the young carriers were unable to recognize the disgusted faces, but they were normal for all the other facial expressions. The non-carriers, who had grown up in households where the disease was present, were able to identify all emotional facial expressions completely normally, which means that the inability to recognize disgust among carriers is not due to social experience.

In this same study, recognition of the noises that signify disgust was also tested, and here no differences were found between those who would eventually fall prey to the illness and those who never would. The HD carrier group and their healthy peers were equally good at figuring out what emotion the sound of retching signified. This suggests that the dismantling of disgust that occurs in HD is progressive, and that the inability to recognize the facial expression of disgust may be a first warning sign for carriers of the HD gene.

In order to determine whether someone was an HD gene carrier, the participants in this study were genetically screened. If you are a carrier of the HD gene, a section of DNA at the end of chromosome 4 has an abnormally high number of base-pair repeats for a specific sequence, "CAG." A healthy person has an average of twenty-three repeats of this CAG sequence. But the mutation that causes HD increases the number of repeats to thirty-five or more. Sprengelmeyer looked at how the mutation might be related to disgust perception and found that the more CAG repeats were in someone's DNA, the worse their disgust recognition was. There is no correlation between the number of CAG repeats and HD severity, but there is a connection with age of onset. The more CAG repeats you have, the younger you'll be when you start developing symptoms. One child, who was diagnosed with HD at the of age three, had eighty-six copies of the CAG sequence.[7]

THE DISGUSTED BRAIN

What is it about having HD that disables disgust recognition? The physical breakdown that occurs in HD is due to progressive deterioration of the basal ganglia, a set of four interconnected brain structures situated at the base of the cerebrum.[8] The basal ganglia primarily control motor ability, but they are also involved in motivation and emotion, especially the emotion of disgust. When healthy people are given the drug scopolamine, which blocks the function of the basal ganglia, their ability to interpret expressions of disgust precipitously drops, and interestingly anger recognition gets worse as well, but they retain normal ability for recognizing faces of fear, happiness, sadness, and surprise.[9] Because HD causes degeneration of the basal ganglia, all the functions that the basal ganglia regulate become compromised.

HD damages more than just the basal ganglia. Another brain structure that is even more intimately involved in our experience of disgust is corrupted as well. Next door to the basal ganglia, neuroanatomically speaking, is a curious part of the brain called the insula. The insula is tucked beneath the temporal, frontal, and parietal lobes.[10] And because it looks like a distinct section of the brain, it was originally thought to be a separate lobe altogether. Currently the insula is categorized as part of the temporal lobe.

The insula is a raunchy, devilish brain region. It is responsible for self-indulgence, sensual pleasure, and the temptations of addiction. Freud would have dubbed the insula home to our id. Besides controlling our hedonistic drives, the insula—and specifically the anterior (front) of the insula—has been shown by numerous brain imaging experiments to be the part of the brain most activated when healthy adults are made to look at pictures of mutilation or overflowing toilets, asked to think about eating cockroaches, or shown disgusted faces. But when

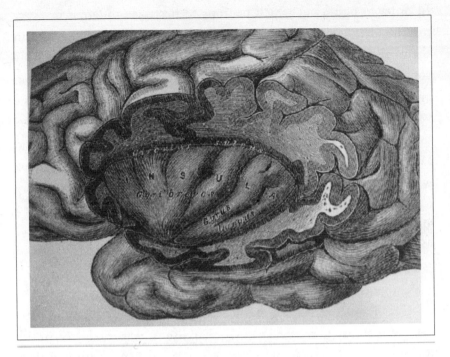

Figure 3.1

The Insula: Original lithograph plate from *Gray's Anatomy,* **1918**

HD patients are given the same kind of pictures to look at and the same emotional prompts, their anterior insula barely glows.[11] Because the insula and basal ganglia are directly connected, diseases that attack the basal ganglia also assault the insula. For example, young adult carriers of HD who are not yet symptomatic show a major reduction in insula volume.[12] Patients with Parkinson's or Wilson's disease, both of which are characterized by a breakdown of mobility and destruction of the basal ganglia and insula, also have great difficulty recognizing the emotional expression of disgust.[13]

The insula even controls the physical side effects of disgust: nausea and vomiting. We know this because of a fascinating tale of patient-guided medicine. Wilder Penfield was Canada's most famous neurosurgeon. In 1950, with his colleague Herbert Jasper, Penfield invented a groundbreaking technique for treating epilepsy in which the specific neurons that triggered a person's seizures were selectively destroyed.

In order to figure out which neurons to destroy—different in every patient—Penfield had to get his patients to tell him what spots in their brain caused their epileptic attacks. While his patients were fully conscious and on the operating table with their skulls sliced open and brains exposed—the brain doesn't feel pain and only a surface analgesic was used to make the incisions—Penfield would apply electric probes to various areas and ask the patient to describe what they were feeling. When the sensation of a seizure began Penfield then accurately knew what specific site to wipe out. In conducting these probes, Penfield discovered a number of specific brain regions that consistently triggered particular responses. Electrically tickling the temporal lobes set off vivid memories, and when he stimulated someone's insula they wanted to vomit.[14] A normally functioning insula enables us to experience both the emotional and physical components of disgust.

The first brain imaging study to identify that the insula was critical to our ability to recognize emotional disgust was conducted in the mid-1990s.[15] Since then, numerous experiments have confirmed that, though some other brain structures are also involved, the insula is home plate for our ability to feel and recognize revulsion. It was a lucky combination of scientific curiosity and opportunity that led to the discovery that HD patients are missing the emotion of disgust and from this breakthrough the brain's inner workings of disgust have been mapped out. It turns out that another affliction, which has nothing to do with physical deterioration or movement, is also abnormal for disgust, but this time in the opposite direction. This psychiatric malady has helped us to understand more about how the brain, as well as our social experience, governs our ability to know when other people are disgusted.

THE ALL-DISGUST DISEASE

Do you wipe off the doorknobs in your home any time someone touches them? Do you count every line marker on the road on your commute to work and if you miss one have to drive back and start over? Are you unable to purge your mind of thoughts of raping your son's teacher, or running over a pedestrian with your car? Do you feel compelled to wash your hands so often that they bleed?

If you've answered yes to any of these questions, then you may be suffering from obsessive-compulsive disorder. Obsessive-compulsive disorder (OCD) is almost as common as asthma and afflicts between 2 and 3 percent of the population. It strikes men and women equally across all social classes and counts many celebrities among its ranks, including one of my favorite actors, Leonardo DiCaprio.[16] OCD is not yet fully understood and no definitive cause has been found, but there is evidence that genetic factors play a role. There is also no known cure, though antidepressants such as Paxil or Zoloft, and behavioral psychotherapy—where the patient learns how to calmly confront obsessions and compulsions—can be successfully used in treatment.

OCD is an anxiety disorder marked by highly intrusive and distressing thoughts and compulsions, and it comes in several forms. Some people are overwhelmed by troubling images, others by the urge to engage in some behavior. Among those who must "do," some people need to check things all the time and are clinically dubbed "checkers," while others are more fanatical about contamination and washing, and are called "washers." Unfortunately, carrying out these rituals, which can include washing, cleaning, checking, touching, counting, arranging, and hoarding, only produces temporary relief. Because there is no long-lasting feeling of satisfaction, these rituals may be repeated ceaselessly. Moreover, people with OCD often believe that if a ritual isn't performed correctly, something dreadful will happen. If you have OCD, it can literally take over your life. Your hands may be perpetually chafed and scabbed, you may be chronically late for appointments because of the number of times you have to recheck that you've locked the front door, or you may simply be unable to leave the house at all because you are so overwhelmed by disturbing thoughts. Indeed, because inappropriate sexual, violent, and blasphemous images are common obsessions in OCD, demonic possession was at one time believed to be the culprit for this disorder, as it was for HD.

OCD shares another dysfunction with HD. Although OCD sufferers frequently over-experience disgust, it turns out that they are quite bad at recognizing when other people are disgusted. Reiner Sprengelmeyer and his colleagues tested OCD patients, patients with Tourette's syndrome, patients with Tourette's and OCD (Tourette's patients often also have OCD), patients with generalized anxiety disorder, and a group of healthy people.[17] As in the study with HD patients, the various groups all performed equivalently on tasks of recognizing age, gender, facematching, and the direction of their gaze. But when asked to identify what emotion the various faces depicted, the patients with OCD performed differently from the rest.

Every patient in the OCD group showed a complete inability to recognize the face of disgust, and every patient with Tourette's who also had OCD showed a specific lack of disgust recognition. By contrast, the ability of these groups to recognize all other facial expressions was normal. The anxiety disorder patients and those who had Tourette's without OCD were good at recognizing all facial expressions. In fact, the anxious patients were slightly better than the healthy participants at recognizing anger, fear, and sadness. OCD and HD share the same odd inability to see disgust. What is more, obsessions and compulsions often occur in patients with HD. The latest research on OCD, however, indicates that rather than every patient with OCD showing a disgust deficit, in order for OCD sufferers to be blind to disgust, their symptoms have to be severe. And the worse their symptoms are, the worse their disgust recognition is.[18]

One might suppose that having certain types of OCD might make you more susceptible to disgust problems, and that "washers"—because they are so consumed by ridding themselves of germs and contaminants—would be the worst at recognizing the face of disgust. This turns out not to be the case.

Kathleen Corcoran, a psychologist at the University of British Columbia, and her colleagues tested patients with various forms of OCD, people who suffered from panic attacks, and healthy adults for their ability to recognize the facial expressions of anger, fear, sadness, and disgust. Corcoran found that the panic-disorder and healthy participants performed very accurately and nearly identically at recognizing these four emotions, but the OCD patients, while good at identifying fear, anger, and sadness, were bad at recognizing disgust. More importantly, not all OCD patients performed alike. Some people diagnosed with OCD were perfectly capable of recognizing disgust while others were not able at all, but it didn't have anything to do with what form of OCD they had—it just mattered how bad their OCD was. When the patient groups were admitted to this experiment, a standard test called the Yale–Brown Obsessive–Compulsive Scale was administered to the OCD group to determine how severe their OCD was. The higher the score on this test, the worse someone's OCD is—the more symptoms they have and the less well they are generally able to function. Corcoran found that the higher someone's admission test score was, the worse their ability to recognize disgust was. OCD patients with low scores were capable of recognizing disgust normally, but people with very high scores had no idea what a yuck-face meant. OCD sufferers exhibit a bizarre disconnect between experiencing and seeing disgust. They are hypersensitive to it, yet can be blind to it in others.

The neurocircuitry of obsessive-compulsive disorder and the brain's processing of disgust are very similar.[19] Both critically involve the insula, just not in the same way. Instead of a deadened response, people with OCD show heightened levels of insula activity compared to healthy adults when they are exposed to pictures of overflowing toilets and mutilation.[20] This explains why having OCD makes one so sensitive to experiencing disgust. It has also been found that OCD patients exhibit elevated metabolism in the primary olfactory cortex, where smells are processed.[21] In fact, it turns out that how sensitive you are to scents is related to how bad your OCD is. When OCD patients were given a difficult odor discrimination task, the *better* they were at the task, the worse their OCD symptoms were.[22] In other words, people with severe OCD are extrasensitive to smells, even though they cannot tell by looking at your face when you've just sniffed spoiled milk.

There is another interesting connection between smell, OCD, and disgust—the brain chemicals that make them tick. GABA[23] is the chief neurotransmitter involved in the sense of smell, and it also courses through one of the key brain regions involved in disgust, the basal ganglia. This is why when we smell something that we don't like, we are often "disgusted" by it. A recent genetic study suggests that a variant in the gene that controls GABA transmission is involved in susceptibility to OCD.[24] As we enter our golden years, our ability to perceive odors weakens and our reactions to disgust diminish as well. It has also been shown that the symptoms of most people who have suffered with OCD for decades either disappear entirely or dramatically improve when they become seniors.[25] Autopsies from the brains of the elderly, regardless of their prior health, indicate that of all neurotransmitters, GABA has wasted away the most. The decline of GABA as we age may therefore explain both our waning disgust and olfactory reactivity, and on the flip side an abnormality in GABA function may cause hypersensitivity to odors and disgust.

Logically, we would predict that heightened sensitivity to disgust, as is the case with OCD, should make OCD sufferers exceptionally good at identifying disgust in other people. So why are they bad at it? From the disorders of disgust so far, we have learned how this emotion is processed in the brain. But in order to properly recognize disgust we need more than just working neural hardware, we need normal social-emotional experience and interactions with others.

To feel disgust or not requires the ability to appraise a situation or object for its possible disgustingness. Through our social development, we learn the facial expressions for various emotions by mapping the expression displayed by other people onto our own emotional states. You and I see a cockroach crawling up the chair leg and we both make a face. I know how I feel and I see your grimace. Through repeated experiences like this in different situations, we come to learn what disgusted faces look like. But if you have OCD you're almost always feeling some sort of disgust, and these frequent, exaggerated, and irrational reactions of revulsion occur when other people around you don't feel disgust at all and hence aren't scrunching up their noses, squinting their eyes, or sticking out their tongues. This leads to a mismatch between your own experience of disgust and what the face of disgust looks like. Leonardo DiCaprio may look at a doorknob and think "yuck," while someone else in the same room has a blank expression or might be smiling or pouting, for whatever reason. Because the person with OCD hasn't experienced many instances where his feelings of disgust are mapped by someone else sharing the same emotion, he has a poor understanding of what it looks like to be disgusted. This social learning may also explain why heroin addicts are, conversely, especially *good* at recognizing when someone else is disgusted.

A recent study conducted at a methadone clinic in London found that heroin addicts were 47 percent more accurate in recognizing the facial expression of disgust than either ex-addicts or non-addicts.[26] The researchers speculated that the reason the current addicts were so good was because they experience so many occasions when people look at them with disgust. Drug users are typically viewed with revulsion by others. Drug users also often feel self-disgust, and this mind-set could prime them to see disgust mirrored in other people. The same explanation might account for why people with generalized anxiety disorder are better at recognizing the expressions of fear, anger, and sadness than psychologically healthy adults.[27] Highly anxious people are internally primed with these emotions and may also experience them more frequently in their dealings with other people. These findings reveal how social cues and social experience are critical for the interpretation and recognition of disgust. In fact, social feedback is so powerful that your disgust *is* my disgust.

DISGUST EMPATHY

Do you remember that fateful day in elementary school when the redhead threw up all over his desk and then a bunch of other kids lost their lunches as well? Maybe you did too? Why did seeing the redhead vomit make you sick to your stomach? The answer is because seeing someone being disgusted makes us disgusted, too.

The neuroscientist Bruno Wicker, along with his colleagues at CNRS (Centre National de la Recherche Scientifique) in France,[28] showed a group of healthy participants videos depicting a model making distinctive facial expressions when he or she smelled disgusting odors, such as vomit and urine. While the participants watched the videos their brains were scanned with functional magnetic resonance imaging (fMRI). Then the participants had their own turn at sniffing the odors and their brains were scanned again. Regardless of whether the participants were looking at the model sniffing urine or they themselves were smelling it, the same area of the brain was activated, the anterior insula.[29] In other words, seeing someone else's disgust triggers the same neurological activity as when we ourselves are disgusted.

It might have felt unpleasant and unfortunate to throw up just because the redhead did, but disgust empathy can save your life. We humans, along with rats and cockroaches, are the world's most successful generalists. Generalists are animals who can successfully exploit any habitat on earth and eat from the full array of the earth's pantry. On the other end of the spectrum, specialists are species that live in restricted and specific habitats and eat only a few types of foods. The panda bear with his exclusive diet of bamboo is one extreme example. Specialists are born knowing what scents to approach and what scents to avoid, but generalists are not.

For generalists, anything hanging on a tree, growing from the ground, scampering across the field, or skimming through the water could be a possible food source—but some of it might also kill us. It is a tricky business figuring out what's good and what's bad for us when any organic substance could be in our food basket. For example, the scent of "poison mushroom" in one locale could be the same scent as "nutritious food" in another. Therefore, generalists have to *learn* what is good and what is bad in the particular habitat we find ourselves in. Fortunately, our olfactory system is wired to learn which foods are poison and which are nutritious very quickly, as illustrated by the somewhat erroneously named phenomenon of "learned taste aversions."

If you have vomited after eating a richly aromatic food—in my case it was pepperoni pizza—you will not want to eat it again. However, it is the aroma, not the taste of the food that will make you avoid it in the future. Ilene Bernstein, a behavioral neuroscientist at the University of Washington in Seattle, showed that taste aversions are really smell aversions when she studied children undergoing chemotherapy treatment (which causes severe nausea).[30] Before their chemo session, the children had eaten an unusual flavor of ice cream dubbed Mapletoff. After their chemo session, the children were offered two different ice creams, Mapletoff again or a new one called Hawaiian Delight. All of the kids refused the Mapletoff but gladly ate the Hawaiian Delight. Both ice creams were equally sweet (taste) and creamy, but the aroma (which produces the flavor) was different. Mapletoff was rejected because it was connected to the terrible nausea of chemo, whereas Hawaiian Delight was accepted because it didn't have any association with sickness. The reason why our food rejection system is so fast and powerful is because these reactions are highly adaptive. If you have been made sick from eating something poisonous, you don't want to keep repeating the mistake until it kills you.

In the natural world, food poisoning is a serious threat for us. The empathic vomit response is beneficial to our survival and is preserved because of its evolutionary success. In our ancestral history, if a fellow hunter-gatherer took a bite out of a decaying water buffalo and vomited we had a better chance of surviving if we vomited too, because we may have eaten some of that water buffalo as well. If, instead of vomiting, our fellow tribesman made a disgusted grimace, we would also feel disgusted and therefore not try the meat and risk becoming deathly ill. Nowadays, your wife sniffs the milk from the refrigerator and makes a face that lets you know whether it's going into the drain instead of into your coffee. If she vomits from sipping the milk it would be advantageous for you to vomit as well, since if she's suffering the consequences of poisoning, your body is well advised to react in kind. Disgust empathy is wired into our brains and into our guts.

To experience disgust, we need an intact network of brain structures, most notably the insula. But we need much more than that. Disgust is the only emotion that has to be learned—what's wrong with picking up a dead skunk?—and it requires complex thinking and interpretation—why is the saliva of a stranger repulsive, but your lover's saliva sensual and your own saliva inconsequential? Disgust also requires an ability to comprehend the prevailing social order—what's wrong (or not) with belching at the dinner table? In other words, disgust entails considerable cognitive and social capacity and involvement. What happens to disgust, then, when you don't have any consideration for the feelings of others, or feel any connection to society?

DEVIANCE AND DISGUST

Jeffrey Dahmer lured men from gay bars to his house for more "partying." His usual ploy was to drug his quarry by spiking their drinks. Then he would strangle or stab his victim to death. In some cases, he drilled holes into their skulls and poured acid inside, leaving his prey alive and in a zombie state for days while he would torture, mutilate, and sexually molest them. After his victims were dead he had anal sex with the cadavers—even as they decayed—and dismembered their bodies with a hacksaw. When the police finally raided Dahmer's house and remanded him from society, they found a freezer full of body parts. Dahmer kept the heads and penises of his victims as trophies, and hacked out their biceps and other muscles and ate them.

This snapshot of America's most notorious serial killer illustrates the very antithesis of humanity and civilized behavior. Overlooking the sheer act of murder, the desire to mutilate and sexually ravage decaying bodies, with the finale of eating the dismembered spoils, is not merely taboo, it shrieks to the core of our repulsion. Dahmer's apparent indifference to the heinousness of his acts makes it all the more appalling. How could someone be this way?

Lionel Dahmer, Jeffrey's father, wrote in *A Father's Story*, regarding the morning when the police told him of Jeffrey's arrest: "So that was what the police really told me in July of 1991. Not that my son was dead, but that something inside of him was dead, that part which should have made him think about the misery he was causing and so draw back from causing it. We call it a "conscience," "being human," or "having a heart" . . . in my son it had either died or had never been alive in the first place."

What Jeffrey was missing was not merely a conscience or "having a heart." Dahmer was a psychopath—a severe and frighteningly "inhuman" disorder. Psychopaths are distinguished by violent sexual and aggressive behaviors as well as severe emotional and interpersonal deficits. Not surprisingly, psychopaths are also deviant in their ability to recognize emotional facial expressions in others. But there is one emotion for which they are especially abnormal.

In 2002, social psychologists David Kosson, Yana Suchy, Andrew Mayer, and John Libby conducted the first and to-date only test of facial expression recognition among a group of criminal psychopaths.[32] Male inmates from a state penitentiary in North Carolina and a jail in northern Illinois were psychiatrically assessed and allocated to either a "psychopath" or "non-psychopath" group. All the criminals were then shown thirty adult male and thirty female Caucasian faces making the expressions for happy, sad, angry, afraid, surprised, and disgusted.

The scores of the inmates were below normal across the board. The overall accuracy score, averaged over the six emotions, for the offenders was 72 percent, whereas the typical score for non-criminal adults is approximately 86 percent. However, not all emotions were equally recognized. Happy and surprised faces were recognized best by both groups; psychopaths and non-psychopaths scored approximately 98 percent for happiness and 90 percent for surprise. Fear, on the other hand, was least well recognized, and correctly identified only about 45 percent of the time by both groups. But for psychopaths, disgust faces were as badly recognized as fear faces, and correctly identified only 48 percent of the time. Psychopaths are often portrayed as distinctively lacking in their ability to feel or see fear, but it seems that they share this deficit with non-psychopathic violent offenders. Where psychopaths are uniquely abnormal is in their inability to understand disgust. Indeed, if psychopaths—the quintessential cold-blooded, antisocial disorder—are unable to perceive disgust, this suggests that recognizing disgust and the ability to feel disgust are at the heart of what it means to be a socially cooperative human being.

The disorders of disgust shed light on the inner workings of this enigmatic emotion. From work with patients and research experimentation, we now know that there are dedicated neural structures, in particular the anterior insula, that are necessary for us to be able to be disgusted and for recognizing disgust in other people, and if these systems malfunction the ability to perceive and experience disgust malfunctions as well. Psychiatric disorders have further illuminated both the neurological and social features of this unusual emotion. The fact that people with severe OCD are blind to disgust underscores the importance of having a normally functioning brain, but it also reveals how social learning is critical for our ability to accurately recognize disgust in other people. It is worth commenting here that the universality of disgust as defined by Paul Ekman's research with the Fore of Papua New Guinea has lately come under some question. A recent study examining the eye movements made when people look at facial expressions of emotion has shown that the eye movements made by East Asians (Chinese and Japanese) are quite different from those made by European Caucasians when looking at the same face, and that Asians tend to make many more errors in categorizing faces of disgust and fear than Europeans do, regardless of the race of the person making the face.[33] This suggests that not only is learning and social understanding involved in our ability to recognize disgust, social and cultural experience is critical for knowing how to look at other people's faces in order to see it.

Chapter 4

GERM WARFARE

Peter was an experienced traveler, but this was his first time in Egypt. He had spent the past few days with tour guides, amazed and awed by the pyramids and the mysteries of the Sphinx and temples along the Nile. Today he decided to meander the streets by himself, to find the real city he always liked to discover on his trips. After a few hours of wandering he found himself in a surprisingly bleak Cairo slum. Dirty, barefoot children rushed at him begging for money. The old and wraith-thin beggars on the street stretched out their arms imploring him. This was unpleasant, but when he turned a corner into a dimly-lit alley, the unpleasantness increased beyond anything he could have imagined. He saw a nearly naked, bony man on his knees, every inch of exposed skin covered with pus and oozing red sores. Flies, which he barely bothered to swat off, pecked at his scabs, and before Peter had time to take it all in, the man lunged at him, grabbed hold of his arm, and shouted for some sort of assistance. Paralyzed by revulsion, Peter was momentarily unable to move. Then he wrenched

his arm free and, holding it as far away from the rest of his body as
he could, he began to run.

Why is Peter so repelled by this beggar? Why is he compelled to hold
his arm away from the rest of his body? What is the connection between
disgust and disease? What does our disgust at disease tell us about this
emotion and about ourselves?

Disease signals the possibility of impending death, and if we get
too close or are touched by the diseased we might catch that impending
death as well. Since the discovery of germ theory, we have learned that
invisible organisms are responsible for disease; though we cannot see,
hear, touch, or usually smell or taste them, germs have become symbolically disgusting and we are preoccupied with eradicating them. But as
we cannot see the basis of disease itself, we are disgusted by what we can
see: rats (vectors of disease), deformity (reminders of disease), pus-filled
sores (evidence of disease).

It is claimed that humans are the only animal with no natural predator, but that's wrong. We have a devastating predator and it's micro-
scopic. Disease is public enemy number one. Twenty of the twenty-five
diseases that have been identified as currently or historically impos-
ing the highest death rate on humans are transmitted through disgust
triggers such as bodily fluids, vermin, and unchaste sex.[1] Feces signal
bacterial infestation, of which cholera is the most deadly form. Saliva is
the transmission route for tuberculosis, which is the most lethal com-
municable disease worldwide; and smallpox, which is in the arsenal
of weapons of mass destruction, is contracted by inhaling droplets of
infected saliva in the air. Blood can kill by transmitting HIV, hepatitis,
and hemorrhagic fever. Sex kills via syphilis, hepatitis, and AIDS. In
2009, AIDS killed 1.8 million people worldwide, and more than 18,000
people in the US still die from AIDS each year.[2] Rats are renowned carriers of deadly parasites, such as the fleas that bore the bubonic plague,
which killed nearly two-thirds of the population of Europe in the four-
teenth century and is still around today. In fact, animals of all kinds can

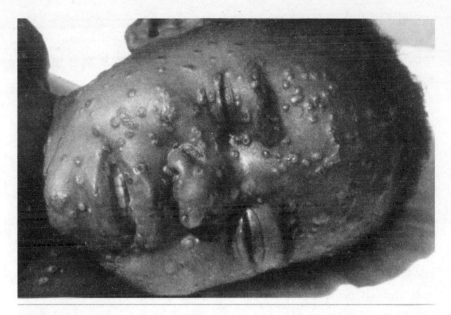

Figure 4.1
Boy With Smallpox

make us sick. Since 1940, there have been 335 new infectious disease outbreaks, of which 75 percent originated in animals.[3] The influenza outbreak of 1918—the original swine flu—killed fifty million people in less than a year and infected one-fifth of the entire world's population. Today, the flu kills at least 36,000 people each year in the US alone.[4]

A SPECIAL KIND OF FEAR

The urge to get away from the hideously diseased, like Peter's reaction to the beggar, is equivalent to fleeing the monsters of our nightmares. But why is the emotion that compels us to flee, disgust, and not fear? The answer is because disgust is a type of a fear—a special type of fear that evolved to help us evade a slow and uncertain death by disease. We are disgusted by oozing scabs, but we fear tigers. Fear is instinctive, automatic, fast, and furious, and helps us avoid death from an urgent danger.[5] The tiger is chasing you—run! By contrast, disgust is learned, cogitative, and comparatively gradual. Even if it feels as though our repulsion at the beggar blistering with lesions comes on the instant we see him, we actually have to take in and interpret the beggar in order to react to him.

The automaticity of fear over disgust was shown in a brain imaging experiment in which facial expressions of fear or disgust were superimposed onto backgrounds that were either the exteriors or interiors of buildings. Participants were asked to indicate either whether the background was outside or inside, or whether the face was of a man or a woman.[6] In other words, their attention was drawn either to the neutral background or to the emotional face. When the face was showing fear, it didn't matter what participants were told to look at—the building or the face; their amygdalas always lit up, just as they would in response to fear. But when the expression on the face was disgust, it was only when the participants were told to focus on the face (to determine gender) that a disgust response was observed and their insulas lit up. Fear is automatic no matter where your mind is, but in order to feel disgusted you have to think and pay attention. Disgust is an unfolding and cognitive emotion; it protects us from creeping dangers that we have to figure out, dangers that are slow in their deadliness, and of which disease, contamination, and decomposition are the foremost threats.

Phobias of blood, vermin, and insects further illustrate the connection between fear and disgust. Needles, rats, and spiders are rarely capable of inflicting direct fatalities. Rather, their ability for destruction comes through the gradual and uncertain path of infection. Interestingly, though spider bites can occasionally make you sick, they very rarely transmit contagious diseases, and as it turns out spider phobia is a Western European trait and not prevalent in non-European cultures.[7] In the Caribbean, spiders are eaten as a delicacy, and in Egypt, spiders are released into a newlywed's bed for good luck. The reason why Westerners loathe spiders is because they were coincidentally connected with disease and death. Spider bites can hurt, and as epidemics of death, such as the bubonic plague, swept through Europe in the Middle Ages and people had no idea why so many of their kith and kin were dying, being painfully bitten by a creepy-crawly offered a good, though inaccurate, explanation. Spiders were also incidentally connected to the real disease carriers, rats, because spiders often nested in the same dark and dank spaces where rats were. However, it wasn't until the nineteenth century that the fleas backpacking on rats were discovered to be the real culprit for the Black Death.[8] Our fear and disgust of spiders was culturally transmitted to us from our European forebears and has infected us psychologically.

Seeing certain types of movement also inclines us toward disgust. Squirming, jiggling, and twitchy motions are more aversive than stillness or straight and smooth movements. The wriggle of the worm or the jerks of the palsy victim are disgusting compared to the smooth strides of a lion or an athlete. Throbbing, spastic, and writhing movements are likely to correspond to sick and contaminating things.

The sight of disease, as in the boy covered with smallpox, can make us very sad, but repulsion is the first emotion we feel. Indeed, the look of sickness is the one disgust trigger that seems to be culturally unvarying. Appearing pale and sickly is at times in vogue—witness pale Victorian languor or the modern-day Goth look—but there is no evidence that any culture has ever found true illness fashionable or attractive.[9] Instead, there is a long history, well before germ theory was ever known, of ostracizing the sick, as in the infamous leper colonies that were rampant during the Middle Ages. It wasn't known how one got sick from being near sick people, but the connection between contact and illness has long been understood. Leper colonies still exist in countries such as Indonesia, but efforts are being made to shut them down.[10]

Chumming up to the person who is covered in oozing sores or choosing to play with rats ridden with fleas is not an immediate death sentence. It can take days or months for disease to complete its mission. An understanding that death can be caused by touching the sick person or by being bitten all that time ago requires complex cognitive reasoning, memory, and abstraction. Fear elicits an instinctive, nonanalytic response to stimuli that can kill us immediately, and we share this emotion with all other mammals. Disgust is different. It is slow, thoughtful, and requires a big brain to be interpreted.

I believe that disgust is the newest and most advanced in the pantheon of the six basic human emotions, and that it evolved uniquely in humans from the emotion of fear, to help us confront our number one predator, pathogens. Disease is slow and uncertain and disgust demands learning and deduction, and humans are the only animal equipped with a sufficiently advanced brain to process this complexity. It is only as mature and socialized adults that we fully come to know the emotion of disgust and understand its implications both physically and symbolically. Young children, primates, and dogs do not know it, do not have it, and do not understand it.

The reason humans needed to evolve the emotion of disgust from the more basic reaction of fear is because we are unique among mammals in how long we can live. The longest recorded life span to date is that of the Frenchwoman Jeanne Calment, who died in 1997 at the age of 122.5. Humans have always had the capacity for very long lives, though without technological resources we don't live nearly as long as we could. For example, someone who was born in Honduras between 1950 and 1955 was projected to live an average of forty-two years, but someone born there between 2000 and 2005 is expected to live an average of seventy-two years. With access to modern medicine and the absence of wars or physical hardship, people born in industrialized countries today have an average life expectancy as high as 87.2 years (women in Japan).[11]

Animals with short life spans don't need the emotion of disgust as much because they are far more likely to meet a fast, fear-based death—by being eaten—than to die slowly—from disease. A Baltimore street rat, which has an average life expectancy of three and a half years, is much more likely to meet its end in an owl's beak or under a tire than to die from a fever. Given our exclusive position of longevity and dominance in the animal kingdom, humans needed an emotional system that could warn us of our foremost killer, pathogens, and therefore we evolved the emotion of disgust. The question is whether another very long-lived cognitively sophisticated mammal, with no predator other than us, experiences disgust as we do. The elephant fits this bill. The elephant is second to us among mammals for longevity, with an average life expectancy of seventy years in the wild, and ethologists from Charles Darwin to the present day have reported that elephants have self-awareness, grieve, and even weep. The conservationist Isabel G. A. Bradshaw, author of *Elephants on the Edge*,[12] observed that elephants appear to have an understanding of death and that grieving rituals are integral to elephant "culture."[13] If it is true that elephants understand their own mortality, then they may very well challenge the uniqueness of human disgust as well.

Whether elephants understand or experience emotional disgust has not yet been tested, so for now I will maintain that humans uniquely possess this emotion. Given the singular human experience of disgust, and that a fundamental purpose of disgust is to help us evade death by disease, a number of intriguing questions arise. Is our personal sensitivity to disgust related to how often we get sick? Does the prevalence of disease where you live influence your disgust sensitivity? Are people who are more vulnerable to illness more easily disgusted? Conversely, do we make mistakes and fail to be disgusted by things that should revolt us if we wanted to avoid disease? Or are we disgusted by things that have no disease threat? And taking this to the next level, if there is such a tangible connection between disgust and disease, could our physical immune system be affected by being emotionally disgusted?

BEING DISGUSTED KEEPS THE DOCTOR AWAY

Imagine that after finishing a delicious dinner at the home of a new acquaintance, your host announces that you've just eaten her pet dog. Chances are high that you would throw up right there, or at least feel extremely nauseated and disgusted. It turns out that this reaction is good for your health, both in the moment and in the near future, because feeling disgusted can make you more resistant to illness. The link is through the neurotransmitter serotonin—one of the chief neurochemicals that's deficient in depression. Besides being involved in a variety of processes, from happiness to sleep, serotonin courses through the stomach and can activate the immune system and trigger nausea and vomiting. Feeling disgusted after being told that you've just eaten your host's dog initiates an immune response; serotonin is released, which stimulates nausea and vomiting—to help you get rid of the unusual meat your body has consumed—and mobilizes your immune system to fight the pathogens that you might have ingested, or any others that might be in the vicinity.[14] For example, if one of the other dinner guests has a cold, you'll be less likely to catch it. The insula seems to be involved here as well. The insula is directly involved in triggering immune responses, especially those that involve a conditioned reaction, as when our headache goes away after taking a pill when in fact all we've swallowed is sugar (the placebo effect) or when we see someone else vomiting and we feel nauseated. The fact that the insula is critically involved in mounting immune reactions against pathogens as well as the neurological underpinnings of emotional and physical disgust deepens the evidence that a fundamental feature of disgust is about protecting us from disease.[15]

Our immune system is so reactive to signals of illness that we don't even need to be physically present for it to kick in. Merely seeing a video of the dog dining debacle could be enough to bolster your disease resistance. For example, in an experiment conducted at the University of British Columbia, healthy participants were injected with a benign bacteria and then shown slides of either frightening images, such as villains brandishing guns, or pictures of people looking feverish, coughing, or pockmarked.[16] Blood was then collected to determine the strength of the participant's immune reaction, and it was found that people who had seen the "sick" slides had a much stronger immune response than people who'd seen the frightening ones. Seeing other people who are ill can trigger our own immune system to mount a response to fight an oncoming infection. Oddly, however, people in this experiment who rated themselves as feeling *more* disgusted by the disease slideshow had a *lower* immune reaction than people who rated themselves as relatively nonplussed by the images. That is, the less disgusting the images were to the person, the more strongly their immune system responded to the bacteria injection. This seems entirely counterintuitive, but perhaps, as with many critical functions, we have built-in backup systems. Our body responds with greater aggressiveness to fight pathogens for us if our mind doesn't prompt us to get away from them by stimulating emotional disgust, and conversely we are more apt to feel disgusted by disease signals if our immune system isn't overly robust. Indeed, as you will see later in this chapter, when our immune system is compromised we are more easily and intensely disgustable than when it is functioning at full capacity. Not only that, regardless of our susceptibility to illness, our general personality predisposition to feeling disgusted can influence how often we get sick.

People who score high on disgust sensitivity, who are particularly contamination-sensitive, and who agree with statements like "I find it difficult to touch an object when I know it has been touched by strangers or by certain people" report fewer colds, stomach bugs, and other infectious ailments than people who score lower on the same measures.[17] Being slightly OCD keeps you healthier. Or, put more generally, the more disgust-sensitive you are, the less likely you are to get sick.

It isn't just our "disgust personality" that influences our health. Several components of our overall personality structure—what psychologists call The Big Five[18]—are also at the root of the connection between staying healthy (or not). One of the Big Five personality traits, conscientiousness—the idealization of the Boy Scout motto "Be prepared"—is linked to better health as a function of disgust.[19] This is because people who are high in conscientiousness are more self-disciplined and prudent, and therefore will be more likely to wash their hands and take general hygiene precautions than people who are haphazard and slipshod—low on conscientiousness. Being careful of mind makes you more careful of body. Conscientiousness can even make you live longer. In *The Longevity Project*, doctors Howard Friedman and Leslie Martin report the results of a major study that followed people from childhood through to their eighties which found that the single best childhood personality predictor of a long and healthy life was con-scientiousness.[20] By contrast, being high on two other Big Five per-sonality traits, openness to experience—which relates to having a vivid imagination, and a general appreciation for art, emotion, adventure, and unconventional ideas, and extroversion—the desire to be with other people, a love of parties, and eagerness to meet strangers—may make you sicker and bring you to your maker sooner, because these traits also increase the likelihood that you'll experiment with potentially risky behaviors. A zealous approach toward trying a new dish in the back-woods of Thailand or having casual sex with someone you just met may be experientially rewarding, but it can put you in the direct path of infection, or worse.

Personality is one of the more heritable human traits. The cor-relation between your personality and that of your mother or father, or between you and your children's personality, is close to 50 percent. Disgust sensitivity correlates at about 40 percent between first-degree relatives, and the correlation for disgust-based phobias, such as spider phobia, is 45 percent. [21]

One might presume that the genetic underpinning of personality would make it less susceptible to environmental influences than, for example, how much garlic you eat. But it turns out that, like cuisine, your personality is directly shaped by the country you live in. In a major cross-cultural study, in which thousands of people from seventy-one geopolitical regions were surveyed, it was discovered that the higher the rate of infectious illness where someone lived, the more "closed" and introverted they were—the opposite of being open to experience and extroverted.[22] That is, the greater the risk of getting sick, the less likely you are to be a party hound or experiment with new foods. This finding could not be accounted for by either religion or income.

The probability of being exposed to an infectious illness is directly related to latitude. The closer you are to the equator, the hotter it is and often the more humid it is as well. Where it's hot and moist, life, from begonias to bacteria, thrives; consequently, these places are host to a large number of infectious pathogens. In Bangladesh, which is 24 degrees north of the equator, the annual death rate from infectious intestinal diseases is 25.3 percent, whereas in Finland, which is just over 60 degrees north of the equator, it is only 0.97 percent.[23]

The fact that geography can influence your personality, however, can still be explained in genetic terms. People who are more careful and socially reserved will be less likely to get sick in disease-ridden locales and are therefore apt to have more children than their extroverted and open compatriots. If their children inherit this guarded tendency, they too will be more likely to reproduce. Thus the inheritance of being introverted and closed to experience is adaptively passed on when it's hot and sticky out.

Your degree of extroversion and openness to experience also relates to your sexual personality. The more open and extroverted you are, the more likely you are to be comfortable with casual sex, having many sexual partners, and seeking sexual variety, whereas the more closed and introverted you are, the fewer sexual partners you will have and the less likely you will be to seek sex outside of a committed relationship. Given the previous discussion, you would expect people who live in hotter, more infectious climes to be less promiscuous than folks from the frozen north, but this turns to be true only half of the time.

Even though people who live in regions with high rates of infectious diseases tend to be more closed and introverted compared to people who live in less germy settings, men across the globe are generally open to experience when it comes to sex. In a study of forty-eight nations ranging from Argentina to Zimbabwe, men by and large had as much casual sex and as many sex partners as they could no matter where they lived, but women didn't.[24] This study used a questionnaire to assess sexual attitudes and personality with questions such as "How many different partners have you had sex with in the past year?" and "How often do (did) you fantasize about having sex with someone other than your current (most recent) dating partner?", and statements such as "Sex without love is okay" and "I can imagine myself being comfortable and enjoying 'casual' sex with different partners." The higher your score on this questionnaire, the more sexually open and promiscuous you are. The data from this transglobal study clearly showed that in hotter countries—that is, in areas with high rates of infectious illness—women were more sexually reserved than their cold-dwelling sisters. For example, women in Bangladesh scored 11.8 on this test, while women in Finland scored 41.6. By contrast, men in Bangladesh scored 31.1, and men in Finland scored a whopping 64—the limiting factor, of course, being the women in the culture with whom the men could have sex. In a similar global study on sexuality and disease prevalence, a strong negative and statistically meaningful correlation between sexual openness and disease rate was found, but only among women—the higher female sexual openness was, the lower the local incidence of disease was.[25] Notably, the results in both these large-scale studies remained the same when national politics, economics, and other sociological factors were accounted for. Together, these findings show that no matter where they live, women are more sexually reserved than men are, and this is because their restraint is biologically adaptive.

SEX GAMES

A simple analogy for the human game of sex is life in a casino with men playing the slots and women at the blackjack table. Genetically speaking, a woman has a card in her hand and is looking for the next card she is dealt to give her a win. A man puts his coins in various slot machines and hopes for payouts wherever he can. Cash for both men and women equals kids, but more importantly, winning means having healthy kids. In the selfish world of genes, the number one goal is to get as many replications of your genes—children—into future generations as possible. There are two ways to do this. Strategy one is have sex with as many opposite-sex partners as you can find. Strategy two is to only have sex with partners with whom you will have healthy and fertile children. Strategy 2 offers the best prospects for genetically high-value offspring—hale and hearty children who will have lots of children themselves, and who will appeal to high-value mates. The reason why health is so important is because if you are sickly, you may not live long enough to be able to reproduce, and even if you do, you are at risk of producing unhealthy children. Moreover, your poor health may make you a less competent parent as well as making you less physically attractive than your healthier peers, which puts you at a disadvantage for attracting mates. Strategy two seems obviously "better" from a genetic value perspective, but it requires much more work than strategy one. Nevertheless, the biological differences between Mars and Venus make strategies one and two more suitable to each, respectively. Strategy one is the slot-machine male approach; strategy two is the blackjack female approach. But the metaphor of life writ Las Vegas is an oversimplification, because the cost–benefit ratio of reproduction for men and women is very unbalanced.

Men have almost nonexistent physical costs in producing offspring, as they need only minutes to engage in the act of sex and have no biological requirement to hang around after that. Moreover, because men have millions of continuously replenishing sperm throughout their lives, they can have a multitude of different sexual partners, each with the biological potential of becoming pregnant, until they are six feet under. Technically, the only limiting factor for a man is the number of hours in the day. Hugh Hefner would be the prototype of success in this regard, though in reality he has only four children.

Women, by contrast, have very high costs to sustain in the pursuit of offspring. In addition to nine months of pregnancy, which requires about a 40 percent increase in energy expenditure and substantial physical vulnerability, there is also a period of at least one year post-delivery when, before the invention of milk formula, a mother could not become pregnant again or she would stop lactating and her infant would starve. Further limitations are that a woman can only become pregnant by one man at a time, she has a limited number of eggs, and a biological clock that ticks down in her forties. In other words, women have to invest approximately two years of their lives and significant physical and psychological resources in order to ensure that the child from only one man will be a healthy baby. Then she has to put in at least another twelve

years of care to ensure that the child itself is reproductively successful. Moreover, she can only do this a limited number of times in her
lifetime. There is, however, one big benefit for women that men don't
have: a woman always knows for certain that the child she is investing
in represents her genetic material and no one else's (surrogate mothers
notwithstanding). Men, on the other hand, can only rely on a woman's
honesty about whether a child is his, which is why lots of mates and
therefore "potential" children is advantageous.

These differences in costs and benefits make women much more
choosy than men about whom to have sex with, no matter how hot it
is outside, and in regions where there is a high risk of infectious disease women need to be especially careful. Unlike blackjack, where the
next card dealt is a mystery, in real life a woman can make deliberate
decisions about what card—that is, man—she's going to take. Another
consequence of the big costs for women and the evolutionary perils of
sickness is that there are particular times in a woman's life when she is
especially sensitive to disease and, as it turns out, disgust.

VULNERABLE STATES

Joan, forty-two, in love and successful, was finally pregnant. But
as the head chef at one of the most popular restaurants in town,
she was now having serious problems. Her high-tech commercial
kitchen was teeming with scents, especially grilled meaty aromas
and savory sauces, and they were making her sick to her stomach.
Joan was disgusted by the smell of roast chicken and flame-broiled
filet mignon, and she couldn't stand the thought of actually taste-
testing any of it (she'd asked her sous-chef to take charge of that).
In fact she could hardly bear to be in the same room with the fresh
steaks, fish, and fowl that her restaurant was famous for. Except for
the frozen meat her husband prepared when she wasn't around,
Joan had cleared her home of everything that wasn't a fruit or
vegetable.

What has happened to Joan, and why? Joan is in her first trimester of pregnancy, a time during which women typically report that their sense of smell has become acutely sensitive. In fact there are no data to back this up. Numerous studies have shown that there are no changes in physical sensitivity to smell during any phase of pregnancy or between pregnant and non-pregnant women.[26] However, there is solid evidence that women's disgust sensitivity becomes heightened during pregnancy, and so does her avoidance of novel foods.[27] Are the real changes in disgust sensitivity and food aversions related to the perceived changes in sensitivity to smells? The answer is yes.

Emotional sensitivity to scents becomes more intense during pregnancy, and emotional sensitivity can feel like physical sensitivity when it comes to odors. Pregnancy is typically a period of life during which women are more emotionally aroused and alert than normal. When we are emotionally involved our attention is piqued, and when we pay attention to scents we become more psychologically sensitive to them. For example, a potent way to make odors emotionally salient and make us pay more attention to them is to advise us that they are dangerous.

In one classic experiment conducted by Pamela Dalton (the olfactory scientist who is testing odor aversion therapy with sex offenders), participants spent twenty minutes in a room scented with a benign, woodsy, balsam fragrance that they were told was either "hazardous," "healthful," or "an experimental standard"—in other words, a neutral odor.[28] To appreciate why the results of this experiment are so remarkable, it is important to know that when we are exposed to an odor for any length of time—two to twenty minutes, depending on the scent—our ability to smell it declines precipitously. This is a biochemical response that occurs in the olfactory receptors in our nose, and it is not under our control. We are all familiar with this experience—for example, when we walk into a bustling Starbucks and are immersed in the rich, heady coffee aromas, but then notice that by the time we reach the counter to order our skinny caramel macchiato we can barely detect the dark, bean-brewing scents in the air around us. To undo this odor numbness we just have to leave the shop and wait a few minutes, and then we can walk back in and appreciate the enticing aromas once again. Participants in Dalton's experiment who were told that the fragrance was "healthful" or "standard" experienced the usual adaptation we have to odors and rated their ability to smell the balsam scent as very weak after twenty minutes of constant exposure. But participants who were told the odor was "harmful" did not. They not only claimed that the scent wasn't any weaker, they stated that they could smell it *more* strongly than they could when they first entered the room. But when they were given a physical test to determine their actual sensitivity to the balsam fragrance, it turned out that their acuity had declined just like everyone else's—just as physiology would lead us to expect.[29] This shows how our emotions, especially anxiety, can amplify our perceived sensation of odors, even though in reality we are no more, and perhaps even less, sensitive to them than we were before the "threat" emerged.

Notably, the scents pregnant women find most intense and aversive are almost always food-related, and meaty as opposed to vegetable or fruit. Meat is an extremely rich source of protein and protein is critical to our survival, especially during the high-energy demands of pregnancy. But flesh also has a relatively high potential to harbor disease-bearing microorganisms. By contrast, though some berries and plants can kill you, rotten fruit and vegetables are much less dangerous than rotted meat. Moreover, poisonous or rotted plants usually taste very bitter and by default we will spit them out. Interestingly, there are two plant aromas that pregnant women often *do* find disgusting: garlic and onion. These plants are unusually high in sulfur, and sulfur is typically an indicator of animal decomposition. Unfamiliar and ethnic foods are also often perceived during pregnancy as unpleasant-smelling, which may be adaptive since novel foods may produce unexpected health consequences. You don't want to try casu marzu for the first time while pregnant, since you don't know how your body will react to live maggots and fermented sheep's milk.

Food and scent aversions are most often reported in the first few months of pregnancy, most especially by first-time moms. The developing fetus is most vulnerable to assaults from inside and outside the mother's body during the first trimester. However, this is the same time when the mother's immune system is most suppressed to minimize her body's rejection of foreign tissue—her fetus. Because of her lowered immune response, if a woman comes into contact with pathogens, she and her fetus are in double trouble. Mom is more likely to get sick, and her fetus is more susceptible to the harmful repercussions. Therefore, it is adaptive for women to be extra-wary during this vulnerable time. And, since meats are more likely than other types of foods to be germ-infested, it is beneficial that they become targets of aversion.

As the balsam scent experiment showed, if your sniff radar is up for possible dangers, everything will seem to smell stronger—and the stronger any sensation is, the more unpleasant it becomes. Your favorite piece of music, if played at a deafening volume, is cacophony, and a caress can become pain when the force is intensified. Likewise, if the aroma of red wine deglazing the steak pan smells exceptionally robust, it can easily become aversive. First pregnancies play a role in odor sensitivity because naïve mothers are especially attentive to, and emotionally involved with, all the vicissitudes of this experience. Foreknowledge enables later pregnancies to be experienced more calmly by our noses and our minds.

CAN'T TOUCH THIS

> When my friend Chris was a graduate student, he would complain that each time his computer was on the fritz and he had to go into the Engineering Department to use the public computers, he'd get a cold. So Chris started bringing a can of Lysol to school with him and would spray away at a keyboard before sitting down to use it. We used to laugh and tease him for this, but Chris didn't get another cold until after he graduated.

What disgusts you more: Your cell phone? A public computer keyboard? An ATM machine? Or a public toilet seat? It turns out that a public toilet seat (not toilet bowl) should revolt you the least if disgust is about avoiding illness, because you're far likelier to catch something nasty, including staphylococci infections which can cause meningitis, or viruses like the swine flu, from your phone, the lunchroom computer, or by using your local ATM than you would be if you licked the toilet seat in a movie theater restroom.[30] In a four-year study that tested surface samples from five major US cities, it was found that computer keyboards were covered with about 3,295 disease-causing bacteria per square inch, which is four hundred times the amount found on the tops of public toilet seats (only 49 germs per square inch), and that cell phones are technological petri dishes. One of the cell phones tested had more than fifty million bacteria on it.[31] Another study conducted in China revealed that *each* key on an outdoor ATM machine held approximately 1,600 germs (keys on indoor ATMs only housed about 965 germs).[32]

People pick their noses, cough into their hands, touch all kinds of dirty things and then punch in their PINs. During flu season, you would be well advised to think twice about using an ATM if you don't have instant access to hand sanitizer or a sink. This is bad news for our health, but could actually have benefits for law enforcement.

The human hand carries about 150 different species of bacteria, which when smeared on various objects can live untended for at least two weeks. The bacteria living on us are highly resistant to environmental stresses like moisture, temperature, and ultraviolet radiation. Moreover, only 13 percent of the various species of bacteria left behind by a sweaty hand swipe are the same between any two people, and you can't get rid of your personal bacteria by hand-washing. Unlike the bacteria you might pick up from borrowing a friend's cell phone, water and soap can't remove our personal bacteria because they are indelibly connected to our genome and live off the by-products of our unique metabolism. This means that a crafty thief who wipes away his fingerprints could still be tracked down by the particular bacterial swirl he leaves behind on the objects he touched, even if the objects aren't discovered for several weeks.[33] The forensic technology to implement this bacterial criminal tracking isn't available yet, but in addition to stocking crime labs of the future, the potential to know what cookie jars we've been dipping into is far-reaching and ominous. Big Brother, not to mention big advertising, could track everything we put our paws on. Criminals who steal money might be in double jeopardy, for getting caught and getting sick—because there is more bacteria on cash than on any other public item.

The filth of money has been with us for as long as we have been handling it. Effectively, exchanging money is like shaking hands with everyone who has ever touched that money, and small bills (especially one-dollar notes) are exchanged many thousands of times in their brief eighteen-month life span. A century ago, a special report in the *New York Times* found that paper money randomly collected from various local stores contained upward of 135,000 bacteria per bill, and all bills were found to have staphylococci living on them.[34] In 1997, another New York City money survey collected coins and bills from hospital workers and found high levels of contamination by both bacteria and parasites.[35] In 2002, a sampling of one-dollar bills in Ohio showed that 94 percent were contaminated with pathogenic bacteria, including antibiotic-resistant staph and strep.[36] More frightening yet, a recent study of paper money in Nigeria found not only contamination by both bacteria and parasites, but showed that the source of the money was key to its contamination level. Food handlers, especially butchers, transmitted more bacteria and parasitic microorganisms than anyone else.[37] Even though this study was done in a less modernized and wealthy country than the US, the implications are just as serious. If you handle dead animals, blood, and money and let it all brew together, your hands become a throng of infestation, which you'll transmit to anything else you touch. Beware the next time you're at the butcher that the person who wraps your T-bone doesn't also make change for you at the cash register. Meat, even from our well-maintained grocery stores, can be lethally dangerous.

WHERE'S THE BEEF?

A story featured in the *New York Times* in the fall of 2009 began by telling the calamitous tale of Stephanie Smith, a twenty-two-year-old dance instructor from Minnesota who is now wheelchair-bound.[38] Stephanie's kidneys could fail at any moment and she has trouble thinking, and though her days are spent in endless hours of physical therapy, she will most likely never walk again. In 2007, Ms. Smith's life was changed forever. Not by a drunk driver or a rock-climbing accident, but by eating one hamburger that her mother cooked for their Sunday dinner, which devastated her nervous system and left her paralyzed from the waist down.

Stephanie Smith's tragedy was the result of an E. coli O157:H7 outbreak that sickened at least 940 people and was eventually traced back to frozen prepackaged patties manufactured by Cargill. This is the same strain of E. coli that killed four children who ate tainted hamburgers at the Jack In The Box restaurant chain in 1994. Later in the fall of 2009, two adults in New Hampshire died from eating ground beef infected with the same deadly E. coli strain. The burgers, which were being sold at stores including Shaw's and Trader Joe's, led to the recall of 540,000 pounds of beef from 3,000 grocers in forty-one states.[39] Every year, one in four Americans will get sick from the food they eat, and about 5,000 of them die;[40] freshly cooked ground beef is to blame for some of the most dreadful outcomes. This news might be enough to make you a vegetarian, but sadly current handling practices can make even the healthiest vegetables dangerous.

According to the Center for Science in the Public Interest, a consumer advocacy group that tracks food safety issues, spinach and lettuce are responsible for 24 percent of non-meat outbreaks of food poisoning, some of which end in long-term disability or death. Witness the E. coli vegetable infestation that killed over 40 people in Europe in the summer of 2011. The causes are contact with animals, unclean water, grimy washing equipment, or unhygienic handling.[41] The moral of this story is that we should revert to a Thoreauvian way of existence to avoid being poisoned by our dinner. A simpler and probably safe enough tactic is to thoroughly wash all store-bought vegetables, buy ground beef only from sellers who grind the sirloin in front of you, and generally know as much as possible about the farm-to-table history of the food on your plate.

If before now, you never worried about eating the spinach soufflé or *steak frites* at your local bistro, you are not alone. Most of us aren't perturbed by the thought of contamination if we don't see it occur or if the contamination is very brief. I admit to invoking "the five-second rule" when a potato chip falls on the floor and I snatch it up before my dog gets to it. Even religion turns a blind eye to low levels of contamination. Under kosher dietary law, you are permitted to eat a food if one-sixtieth of it (or less) has been contaminated with forbidden products; the tap water in New York City was recently the subject of a test of this edict that for a while padded the bank accounts of water-filter manufacturers and plumbers.

There is a tiny creature known as a copepod in New York's drinking water. Copepods are found in water all over the world and are perfectly harmless. But they are a distant cousin of shrimp and lobster, whose consumption violates the biblical prohibition against eating shellfish. Between the spring of 2003 and the fall of 2004, the Orthodox rabbinical authorities in New York and Israel heatedly debated about whether the presence of this crustacean violated kosher dietary law. Eventually, the ruling came down in favor of the minuscule shellfish, with the argument that since it couldn't be seen by the naked eye and since the Talmud was written before microscopes were invented, Orthodox forebears wouldn't have known of its existence. The copepod squeaked in under the radar, the issue was dropped, and the industry of home water purifying went back to business as usual.

We are willing to let various forms of contamination slide by, and we are also unfazed by some dangerous "disgusting" stimuli, but our ill-advised behaviors and human inconsistencies are not limited to the realm of disgust. Our fear response fails us under many sinister situations as well. For example, seeing a snake elicits an automatic fear reaction in primates, including us, but seeing an electric socket does not.[42] Yet the chances of a fatal mishap with a snake are much less than with an electrical outlet. Your house is teeming with them and they're always alive and ready to strike with a potentially deadly dose of 120 volts of electric current. Sticking a coat hanger into one of these innocent-looking fixtures could be lethal. Yet we are blithely unafraid. Why?

The argument is that these newbies haven't been around long enough for us to have evolved the evolutionary advantage to avoid them. Historically, people with the genetic predisposition to be fearful of snakes fared better and were more likely to pass on their snake-avoidance genes than those who thought snakes would make cute house pets. There hasn't been enough time, nor does it seem like there ever will be, as technological gadgetry changes at blinding speed, for individuals with a genetic mutation that causes fear of electric sockets to outlive and outreproduce those without it. Likewise, there hasn't been enough evolutionary time for the genetic advantage of being disgusted by cell phones or dollar bills to be manifested and consequently become a heritable trait.

It isn't our fault that we can't quite manage the threats of today. We are using 50,000-year-old hardware to manage in a twenty-first-century world.[43] No period in human history has ever witnessed such a velocity of technological innovation, and obsolescence, as now. Your grandmother could rely on her grandmother to teach her how to do almost anything. Now we rely on our kids to teach us. Could you have even imagined an iPad fifteen years ago? Does anyone still own a transistor radio? Our instincts come from the Paleolithic, but since public computers are hazards now, we have to use active thought rather than relying on intuition to protect us. Disgust is an instinct that has to be learned.[44] Even so, it seems we have not learned it to our full advantage, if the principal feature of disgust is to protect us from disease.

The devastating track record of malaria is a prime example. It is estimated that half of the people who have ever lived on earth have died from the contagious parasitic disease malaria, and it continues to kill close to one million people each year.[45] Malaria is uniquely transported and inflicted upon us by the mosquito, but we are not disgusted by mosquitoes. Unlike spiders, which we are repelled by and which are blamed for illnesses that they do not cause, mosquitoes are generally dismissed—yet they are the cause of the most deadly scourge in human existence. Why do we make such a huge disgust mistake? There is no clear-cut answer, but some explanation comes from the fact that mosquitoes fly and buzz.

We do not seem to be disgusted by insects that locomote quickly in the air compared to those which creep and crawl on the ground. The classic of all buzzing, flying insects is the bee, and we are not disgusted by them at all; in fact, the bee is often a symbol for cuteness. When we aren't smiling at the cartoon bees on our honey jars, we all tend to be afraid of the real thing flying by our face. This is for good reason. When bees do touch us, they tend to sting, and the sting can be very painful, even fatal for some people. Other flying, buzzing bugs, like flies, typically irritate or annoy us, but they also don't usually disgust us. It seems that the signals of "fly and buzz" elicit fear or annoyance, and the signals of "creep and crawl" instigate disgust—probably because of their slithery movements and because they tend to be slimy and mucusy. Mosquitoes elicit annoyance, and maybe fear if we have reason to suspect malaria, or West Nile virus, but they don't disgust us as they should if disgust is the signal for a disease threat. This "human error" shows that the connection between disease and disgust has to be learned and that even when the threat of disease is intellectually well understood, we are not necessarily propelled to feel disgust—which means, as you will soon see, that disgust is not only about disease.

We encounter many stimuli which should inspire our repulsion if the motive for disgust is to keep disease away, but which fail to make us respond in this way. Similarly, there are various signals that we find disgusting but which are in fact harmless, and in some cases we are disgusted by signals that are actually indicative of reproductive value. Semen is one of them; another is acne. A bad case of acne is a pus-filled, red, and scabby sore bonanza, and it has destroyed the self-esteem and social life of many a teenager. But in spite of the fact that teens and others find the "look" of acne disgusting, it is a hormonal sign of sexual maturity and a green light for the ultimate goal of our genes—to go forth and multiply.

Our overreaction to possible germs is not necessarily to our advantage. Not only can it make people miserable, it can backfire on our health. We have ten times more visiting microbes in our bodies than human cells, and there are about seventy-five trillion human body cells. Many of the bacteria that live inside of us are very friendly. The probiotics, like acidophilus, help us digest our food, and bacteria in our noses fight off pathogens in the air we breathe. In the future, the right bacteria may be used to cure cancer.[46] The bacteria within us even regulate our development. Mice who are reared without any germs in their bodies never fully develop their intestines. In fact without the right bacteria taking up residence inside of us, we would die. In one recent case, a woman with a horrendous stomach infection (*Clostridium difficile*) that causes a constant state of diarrhea had lost sixty pounds in eight months and was wheelchair-bound in diapers. All the conventional antibiotic therapies had failed so, out of options, her gastroenterologist tried an uncommon technique known as fecal implantation. He took a sample of her husband's stool, mixed it in a saline solution, and delivered it into her colon. Her husband's microbes took over and her intestinal flora became normalized. The patient's diarrhea vanished in twenty-four hours and she completely recovered a few days later, with no relapse to date.[47] It's a good thing that disgust didn't prevent either the invention or the application of this treatment, or this woman and patients like her would die.

OF MEAT AND MAN

Eating the meat of specific animals is taboo in many cultures. Non-vegetarian Hindus will never eat beef, observant Jews and Muslims will never eat pork, and most Somalis won't eat fish. Although religious reasons are given for these avoidances, they have some basis in genuine health hazards, such as trichinosis in undercooked pig meat. Noticeably, there are far fewer cultural proscriptions against consuming specific fruits and vegetables,[48] and where there are vegetable prohibitions they are most commonly about garlic and onion—the sulfurous, animal-decay-reminding variety. As the story of Joan showed, pregnancy, especially the first trimester, is a time of heightened alertness and aversions to meats and novel foods. This ends up being advantageous because of the possible dangers that cavalier consumption could produce. However, it turns out that during the first few months of pregnancy, women aren't just extra wary of meat and exotic cuisine, they're also more wary of exotic people.

In one recent experiment, 206 US citizens at different stages of pregnancy, as well as non-pregnant women, read two essays purportedly written by students. In fact, the essays were made up by the experimenters. One essay was supposedly "written by a foreigner" who was critical of American values, and the other was a pro-USA essay "written by an American."[49] The women then rated the essay authors on several scales such as intelligence, likability, and morality, as well as how much they would like to work with each author. The results showed that women in the earliest weeks of pregnancy were much more negatively predisposed toward the "foreign author" and more positive about the "American author" than women in later stages of pregnancy, and they were drastically more biased than non-pregnant women. Women in their first trimester were six times more prejudiced against the foreign author than women who weren't expecting. An increase in sensitivity to disgust, meat, and unusual foods during the first few months of pregnancy makes evolutionary sense. It helps protect the mother-to-be from harming herself and her unborn child. However, when this avoidance of the unfamiliar or foreign extends into the interpersonal realm, it may not be so adaptive. The next chapter reveals just how devastating for humanity it can be when we let disgust manipulate our social decisions and the way we treat other people.

Chapter 5

DISGUST IS OTHER PEOPLE

Helen hurried into her local superstore on Friday afternoon to pick up groceries for the weekend cookout. Her eyes scanned the aisles as she calculated the speediest route while her body moved toward the first lane of carts by the door. She yanked a cart from its convoy, and spinning it in the direction of the hot dog buns, suddenly stopped and grimaced. An "eww" escaped under her breath. A crumpled shopping list lay at the bottom of her cart. Helen let go of the cart and quickly searched for one devoid of remnants from prior shoppers. Spotting a pristine-looking cart, she grabbed it and sprinted on her way.

Why does a crumpled piece of paper make a shopping cart disgusting? The cart that Helen ended up using for her grocery mission could have been touched by hundreds of patrons since its last cleaning and the cart she rejected could have been washed that morning. But a wrinkly paper with a stranger's scrawl is an unmistakable sign that other people have used and touched that cart, and this makes it repellent. Signs of the touch of strangers, whether past or present, are unpleasant to us because they implicitly conjure thoughts of unknown contamination, regardless of the cleanliness reality. Even photographs of crowded places are more disgusting than the same location when desolate.

At the end of a BBC documentary called *Human Instincts* that aired in the UK in October 2002, viewers saw an ad encouraging them to visit a website (www.bbc.co.uk/science/humanbody/mind/disgust) and take a survey. More than 77,000 people from 165 countries went to the website and completed the online investigation, which involved looking at twenty photos and rating how disgusting each one was. Among the photos were pairs of identical pictures, except that one of the pair was intended to conjure thoughts of disease and the other was not—for example, a white towel with a blue stain or a white towel with a reddish-yellow stain. One photo pair was merely a subway car that was either packed with commuters or empty. Overall, more than 98 percent of people rated the disease-suggestive photo as equally or more disgusting than its undiseased counterpart; both the towel with the reddish-yellow stain and the packed subway car were rated as twice as disgusting as the towel with the blue stain or the empty subway car.[1] The colors of blood and pus are understandably linked to thoughts of illness and contamination. But why do thirty people seem dirtier than three?

The answer is because thirty people are a crowd, and the crowd is a swarm. "Swarms" are another general elicitor of disgust for us. A swarm of ants scuttling across your kitchen floor is repulsive, while two ants meandering by can be tolerated. The swarm is vibrant with life, but it is also overwhelming, disorderly, chaotic, and uncontrolled.[2] The swarm has more potential to infect us than the few, because the more there are of any living thing out there, the more chance there is that they will get onto you or into you, and, as mentioned before, a primary objective of disgust is to keep the outside away from our inside.[3] Nevertheless, what or whom "they" are is key when it comes to our concerns about protecting our insides from the outside.

In the same BBC online survey, this time with nearly 31,000 respondents, people voted the mailman as the person they would least

like to share a toothbrush with and their spouse as the most acceptable toothbrush-swapper; the other choices were "your boss," "your
best friend," "the weatherman," and "a sibling"—best friend and sibling
followed closely behind spouse.[4] Sharing bodily fluids with unfamiliar people is more disgusting because strangers pose more of a health
threat, since we don't know where they've been or what they've been
doing. The cooties they carry are mysterious in quality and quantity and
therefore present a threat. By contrast, you already share so much with
family members and other intimates that their germs are less likely to
harm you. This is because you have either already developed an immune
response to their germs, or you carry these bugs around on you too.
However, how much we like the person and how attractive they are also
moderates our desire to swap toothbrushes regardless of how well we
actually know them. In the BBC study, the choice "your boss" came in
second (behind mailman) as least acceptable to share a toothbrush with,
whereas "the weatherman" was three times better than "your boss." Pre-
sumably one has more physical contact with one's boss than with a TV
weather reporter, but bosses are often disliked and TV anchors are usually well-dressed and attractive. As you will see in chapter 7, our sexual
attraction to someone can override all qualms about disgusting body
fluids. Besides mitigating our aversion toward the body substances of
others, our revulsion at the prospect of contamination from other people
transfers to an alternate realm that is literally magical.

BLACK MAGIC

In the August 2009 issue of *Harper's Magazine*, the "Findings" section reported that most people, even if they urgently needed a heart
transplant, would not want to receive the heart of a murderer.[5] Why
are we so irrational? Don't we know that we can't catch a murderer's
personality from one of his internal organs? Although there are differences in the degree to which each of us feels this way, nearly all of us

experience an eerie discomfort at the thought of physical contact with something that was owned by an infamous or "evil" person. We are somehow worried that we may become possessed or tainted by their nefariousness. This predisposition to fear invasion by wicked souls is due to an implicit belief in "sympathetic magic" or "magical contagion." Sympathetic magic is considered to be a basic and universal principle of human thinking[6] and it is overtly acknowledged in many traditional societies. The Hua of New Guinea believe that a person's vital essence resides in what he or she wears and works on, and that, for example, the produce from a farmer's garden is infused with the essence of his soul and will be tasty or tart accordingly.[7] In cultures such as ours, where we don't explicitly accept such beliefs, the secret feeling that the spiritual essence of a person or thing can be transmitted to us through physical contact is still persuasive enough that even the most rational among us find it nearly impossible to resist.

In multiple experiments, Paul Rozin from the University of Pennsylvania and his collaborators at various institutions have demonstrated people's repulsion at the possibility of "catching" someone's or something's spiritual essence.[8] In these experiments, participants are asked about scenarios such as whether they would wear a sweater that belonged to Adolf Hitler after it had been dry-cleaned and sanitized, drive a car that had been previously owned by a murderer, or drink their favorite juice after a dead and sterilized cockroach had been dipped into it. In all the scenarios, most people refuse the "contaminated" objects and state that the prospects make them feel icky and creepy. Yet when asked what is wrong with the items, most people have a hard time explaining why they feel so uncomfortable because they realize the illogic of any justification.

Of course, the ghost of Hitler's evilness is not residing in his sweater ready to possess you the minute you put it on. Yet the feeling is still there, and many experiments have verified the effect. People don't want to sleep in a hotel bed if they find out someone has died there, or use silverware that was owned by someone with cancer, or swim in pools where psychiatric patients have swam (none of these conditions are contagious).[9] We are even reluctant to borrow a car from someone who has suffered a misfortune or accident through no fault of their own, such as getting AIDS from a blood transfusion.[10]

The irrational yet unshakable rule we feel is "once in contact, always in contact." If something was once in contact with something or someone bad, then that badness is forever in it. Hitler's possessions are indelibly stained by his evil persona. The cancer patient's misfortune stays lurking in her cutlery. The psychiatric patients infect the swimming pool with their mental illness. And this irremovable evilness, illness, or misfortune can be transmitted to anyone who makes contact with it. Inanimate objects with "disgusting" usages also forever make them objectionable—even being close to them can turn something desirable into something disgusting. People are reluctant to take chocolate-chip cookies from a sealed package if the package is placed next to feminine hygiene pads on a store shelf. If the two items make slight contact, as they might in a shopping cart, the level of repulsion jacks up a few notches.[11]

Another type of sympathetic magic is based on how something looks. It if looks like a duck, it must be a duck. Form is function.

> Sean's gourmet baker friend, Fenton, offered to make him a special cake as his fortieth-birthday present. Sean was delighted. He knew that his friend's cakes were showcased at the finest parties and restaurants in town, and he'd sampled some of these delicious confections before. What Sean didn't know was that Fenton had a perverse sense of humor. The day arrived and, with twenty friends in attendance, Fenton sailed into Sean's house with the pièce de resistance.

The photograph on the next page is the cake Fenton made for Sean. It is a real cake, made from sugar, flour, butter, eggs, chocolate, cream, vanilla, food coloring, and a little cardboard for structure. But, as you

Figure 5.1
Feces in Toilet Birthday Cake

can imagine, Sean was a little flustered and dismayed by his friend's humor. Fortunately most of the guests thought the cake was a good joke and ate some after the most daring had cut in. But would you be among the first to take a slice?

Paul Rozin tested a similar predicament by asking people to rate how much they'd like to eat chocolate fudge shaped like dog feces, hold a clean rubber sink stopper that looked like vomit between their lips, or drink apple juice from a brand new bedpan. Just as with Sean's cake, people felt repulsed by the thought of putting fudge, rubber, and apple juice into their mouths if it *looked* like something disgusting. This is the sympathetic magic rule of similarity. If something looks like urine, vomit, or feces, however superficially, we worry that it might actually *be* what it looks like. A cake that looks like a toilet bowl containing an unflushed bowel movement must have the aura of real human waste in it. Even food that accidentally resembles human body products, such as pea soup, can make an unadventurous eater jumpy.

This human proclivity for the sympathetic magic of similarity is in the same constellation of inclinations that make us naturally superstitious and that make voodoo so intuitively scary and persuasive. If you stick a pin in a doll depicting your adversary, it *feels* like you might be hurting the real person. I became quite disturbed one afternoon in a convivial bar on Martha's Vineyard when I noticed that between the bottles of tequilas, cognacs, and gins a little cloth doll of Marge Simpson was perched on the bar shelf—with hundreds of pins stabbing into her. This was a doll of a cartoon character—the unreal of the unreal—and still I felt distressed. I like Marge Simpson.

The evolutionary explanation for our magical inclinations is that our ancestors who reacted more vigilantly to "strange" signs outlived those who didn't pay attention to whether the rustle in the leaves was a ghost, the wind, or the enemy stalking you. Likewise, our ancestors who paid attention to symbols that portended their destruction and took heed survived better than those who didn't care if their effigy was made into a pincushion.

PRETTY AND POPULAR

Positive sympathetic magic exists as well. We will readily believe that the good and interesting essences of other people can be transmitted through things they have once touched, owned, or which were physically part of them. I knew a physicist who told me that after she had a kidney transplant she unexpectedly developed a craving for pistachio ice cream. She later found out that pistachio ice cream was her kidney donor's favorite dessert, and despite her rational profession she would routinely declare that her new found ice cream fondness came from her new kidney.

Good essence can even undo evil essence, but only up to a point.

Bad trumps good in all life experiences. For instance, in all my years of manipulating participants' moods in the laboratory, I have always found it easier to make someone feel noticeably anxious than to make them feel particularly good. This isn't just my poor choice of happy-induction tactics; it's a widespread effect and the reason why most emotion experiments involve negative rather than positive mood. It just works better. It may not feel so, but the imbalance of bad over good is adaptive. Avoiding bad things gives us much more of a survival advantage than approaching good things does. The tiger will eat you and end you, whereas your sexy and magnanimous companion won't necessarily save you or mean that you will go forth and multiply.

For her dissertation research, one of Paul Rozin's students, Carol Nemeroff, who is now a professor at Arizona State University, asked people to imagine a brand new sweater that had come into contact with various bad people or things, such as Hitler, dog poop, and hepatitis, and then to imagine how various manipulations, such as sterilizing the sweater or having it worn by a good person like Mother Teresa, could renew the sweater's quality.[12] Interestingly, sterilizing worked to cleanse the sweater of physical nastiness, such as hepatitis, but it didn't work to purify it if Hitler had worn it. Only Mother Teresa donning Hitler's sweater could diminish its evilness, but she couldn't take the tarnish off completely. The only manipulation that could fully eliminate Hitler's evil essence was to destroy the sweater completely—by burning. Burning has traditionally been a method for obliterating supernatural evil, such as in the unfortunate era of witch hunts, or even in the act of burning books whose contents spiritually threaten us. Burning, both literally and symbolically, reduces evil to ashes.

There is a situation, however, where positive physical contagion is particularly compelling. Several experiments in shopping settings have shown that when a shopper believes that an extremely attractive person of the opposite sex has recently touched an item, the amount that shopper is willing to pay for the item increases substantially. This effect is especially pronounced when the highly attractive "toucher" is compared to someone of only average looks—and especially if the shopper is a heterosexual man. In an experiment conducted at a college campus store, the amount that men were willing to pay for a T-shirt leaped to $19.71 when a beautiful woman walked out of the dressing room and put it back on the rack—from a measly $7.44 when a woman of only average attractiveness brought it out of the dressing room.[13] When no one had touched the T-shirt, the accepted purchase price was about $15. This is why advertisers use beautiful people to sell us nearly everything, from cars to mobile phone plans. Beauty isn't the only thing that gives you monetary marketing power. If you're idolized, your essence has enormous capital as well. John Lennon's white suit, worn on the *Abbey Road* album cover, sold for $118,000 in a 2005 Las Vegas auction.

Our reaction to the magical contagion that spiritual essences can impart to objects is an overextension of the adaptive response to get away from things which may be diseased and, vice versa, to make contact with health and power. These superstitious beliefs and behaviors are generally harmless and most likely have their origins in reactions to "uncanny" situations that were beneficial to our ancestors. Before germ theory was proven to be a medical reality, the belief that essences of evil or death could be transmitted from one person to another via the things they touched would be advantageous. If you sleep on a blanket that someone who just died was last using, you can indeed catch what killed them, as was the case in the smallpox tragedy that killed at least 60,000 Native Americans. Unlike the hotel bed scenario, the Native Americans didn't have any choice but to take these blankets, and it is argued that the colonizing Europeans who donated them were committing a calculated act of genocide. This is a direct example of disease being used to contaminate an object with a real and intended physical consequence for those who make contact with it, and it is a variant of germ warfare. More frequent and more sinister, however, is the use of the unreal and indirect sympathetic magic of similarity and contagion to incite detestable behavior.

BAD BEHAVIOR

In 1925, Hitler wrote in *Mein Kampf*, the monograph which laid out the blueprint for his future campaign of extermination, that Jews were "maggots in a festering abscess" who were set on destroying the clean body of the German nation. When Hitler became chancellor of Germany in 1933, he established a Ministry of Propaganda with Joseph Goebbels at the helm. Under Goebbels's directorship, the Nazis took over the media and began radio broadcasts and print propaganda that depicted Jews as literal disease-mongering parasites. The picture on the next page is a German propaganda poster that hung over the streets in Nazi-occupied Poland and says "Jews-Sucking Louse-Typhus."

In 1940, the Ministry of Propaganda established a department of film and that same year produced *Der Ewige Jude* ("The Eternal Jew"), the most famous Nazi propaganda film ever made. The film is a mixture of documentary footage from the Nazi invasion of Poland and studio-filmed scenes that depict Polish Jews as corrupt, filthy, lazy, ugly, perverse, and barbaric. In particular, Jews were likened to hordes of rats scrabbling out of filthy sewers, spreading their power and beliefs like a virulent sickness. Even medical texts at the time described Jews as "fungoid growths," bacteria, tumors, and cancer.[14] In a characteristic 1936 lecture to medical students, the SS radiologist Professor Dr. Hans Holfelder showed a slide in which cancer cells were portrayed as Jews and, not surprisingly, the cancer-fighting radiation as Nazi storm troopers.[15] Not only did the Nazis brand Jews as pathogens, they accused them of trafficking in disease-causing products, such as tobacco. The take-home message: stamp out Jewish vermin and infection. The message grievously succeeded.

Another notoriously verminized group were the Tutsi in Rwanda. In April 1994, Rwanda descended into a hundred-day nightmare of unrestrained violence during which nearly one million men, women, and children were slaughtered. A year earlier, Hutu hardliners had begun broadcasting "exterminate the Tutsi cockroaches" over the airwaves in

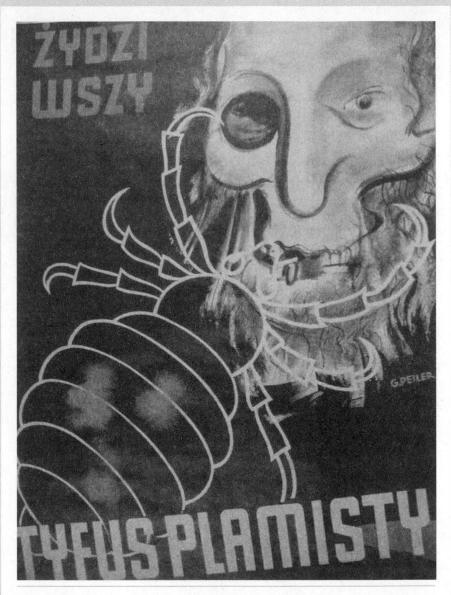

Figure 5.2

Nazi propaganda poster from Poland, 1942

order to galvanize Hutu militants, as well as desperate and frightened Hutu citizens, into committing this massacre. The "final solution" is frighteningly simple. If you want to make a group despicable and justify murdering them, equate them with disease and disgust. If a group of people are analogized as vermin they take on the essence of that vermin, making it easier on our conscience to destroy them.

Racism against immigrant and ethnic groups in the United States also has a disgraceful history of denigration via disease. Since 1880, the US has seen waves of foreign newcomers, and each time the "aliens" are greeted with less than open arms and rumors that they pose threats to the health of rightful citizens. The "open air school movement," which originated in Berlin in 1904 and won popularity in North America after World War I, was a widespread public health initiative that was so influential it had a major impact on school architecture. The purpose of the open air movement was to clear and clean the air of tuberculosis germs that immigrant schoolchildren, predominantly Italians and Jews, were suspected of spreading.[16] Classrooms were built with walls of enormous windows that could be fully opened and kept open all the time, which indeed they were even in frigid winter months and snowstorms. This itself might have exacerbated respiratory illnesses. I was intrigued to discover that, coincidentally, the first open air school in the US opened in the dead of winter, 1908—just a few miles from me—in Providence, Rhode Island.[17]

Hollywood assisted in disseminating propaganda about disease-spreading foreigners in films like *Panic in the Streets* (1950). Set in humid New Orleans, Richard Widmark plays an American doctor who has forty-eight hours to defeat an outbreak of pneumonic plague that is being spread by the character Blackie, a low-class, gambling foreigner variously described as Armenian, Argentinian, and Greek. Widmark triumphs at the film's end when, after a lengthy chase sequence through dockyards, Blackie is trapped by a rat-catcher on a ship's mooring line and, already injured by a gunshot, falls into the water where, true to the rat metaphor that has been conjured for him, he drowns.

In more recent history, immigrants were indiscriminately designated as "high risk" for bringing AIDS into America. Between 1990 and 1993, the Immigration and Naturalization Service (INS) quarantined HIV-positive Haitian immigrants at Guantánamo Bay—in prisoner's cells. During the same period, however, HIV-positive foreign *tourists* were allowed full entry into the US, to spend their money and then leave. On June 10, 1993, President Bill Clinton signed a law stipulating HIV infection as a criterion that would bar immigrants from obtaining US residency status. The reasoning behind the law was supposedly to protect US citizens from disease exposure and the added expenses of providing medical care to foreigners. California's Proposition 187, which was enacted in 1994, was even more severe and required publicly funded health-care facilities to refuse care to illegal immigrants and to report anyone who met such a description to the INS. Of course, this only made sick illegal immigrants less likely to go to doctors. Proposition 187 was later found to be unconstitutional and was terminated by Governor Gray Davis in 1999. Nevertheless, anti-immigration rulings in states like Arizona could bring about the same problems. The irony of these laws is that they do not make immigrants leave; rather, they increase the likelihood that contagious illnesses will go untreated and consequently spread, because intimidated foreigners won't seek medical treatment. This not only increases the risk of infection to US citizens, it increases the fiscal burden of health care since having lots of sick people—foreigners and citizens alike—means higher health-care costs. What makes the foreigner-disease-transmisson belief even more stunning is that it has been established, though not much publicized, that the number one route of pathogen introduction to North America is not via immigrants or refugees, but via military personnel.[18]

Military personnel who have been posted to various hot and chaotic climes do not undergo any medical screening if they don't complain of sickness prior to returning home. Typhoid fever and hepatitis are regularly brought into the US by soldiers who return from being stationed in Afghanistan and Iraq. By contrast, immigrants who wish to enter the country legally must undergo extensive health screening and testing before being allowed to cross our borders.

Travelers returning from abroad are also disease vectors, and they don't get screened upon reentry either. The 2003 Canadian SARS epidemic originated when a Chinese–Canadian returned home to Toronto after visiting Beijing. By the end of 2003, forty-four people in Toronto had died of the disease. The worldwide SARS death toll that year was 800, including 350 in China. (No one in the US died of SARS.) Swine flu and other virulent viruses are also trafficked by homecoming travelers much more so than they are by foreigners. For example, the 2009 swine flu epidemic in the US was chiefly spread by tourists returning from trips to Mexico.

FAT BUT NOT HAPPY

Immigrants, foreigners, and religious and ethnic minorities aren't the only social pariahs. The less lean and beautiful among us are perceived as disgusting and the perpetrators of disease too. Obese people—anyone with a body mass index (BMI) of 30 or more, such as a five-foot-four-inch woman weighing 174 pounds or a five-foot-ten-inch man weighing 209 pounds—routinely suffer social discrimination and abuse, and are outright targets of disgust.[19] The obese are stereotyped as dirty, sweaty, and smelly as well as unintelligent, unsuccessful, messy, lacking will-power, and lazy. College students claim that they would rather marry an embezzler, a cocaine user, or a blind person than someone fat.[20] Shockingly, our own body weight doesn't affect our aversion for fat people. Fat and thin adults alike rate fat people as disgusting.[21]

Obliviousness to our own girth may partly explain why US obesity rates continue to soar. In 2009, over 33 percent of all adult Americans were classified as obese, compared to only 15 percent in 1980.[22] Childhood obesity is also growing at an alarming rate. The child and teen obesity rate is at 30 percent or higher in thirty states, with Mississippi having the greatest percentage of obese and overweight children, at 44.4 percent. Nationally, 18 percent of adolescents aged twelve to nineteen, and 20 percent of children aged six to eleven, are obese.[23] This is triple the number of obese children that was recorded in the US in 1979. Even the animals that live around us are getting fatter. The average weight of pet cats has increased 10 percent each decade since 1960, and dogs have gotten fatter by 3 percent in the same time span. Even Baltimore street rats have chubbed up 7 percent each decade over the past fifty years.[24]

In addition to derision, social exclusion, and heightened risk for cardiovascular and metabolic diseases, there is a considerable financial cost to being overweight. Obese women earn 6 percent less than women of average weight doing the same job, and obese men earn 3 percent less than their leaner counterparts. Over a lifetime of employment, this could mean lost earnings in the six-figures.[25] On-the-job discrimination is another serious issue. The obese are thirty-seven times more likely than people of normal weight to report weight-related employment discrimination, and severely obese people (BMI above 35) are more than one hundred times more likely.[26]

The negativity of being fat in the workplace is even contagious. In one study, not only was an obese job candidate judged to be less competent and worthy than a person of average weight, but a man of average weight sitting next to a professionally dressed obese woman was judged to be a much less capable and desirable job candidate than if he were seated beside a similarly dressed woman of average weight. Ironically, the female job candidate was the same person in both conditions, either wearing a fat suit or not.[27] The negative contagion of being near a fat person also persisted when it was made clear that the physical proximity was accidental. That is, even when the judges explicitly knew that the job applicants were complete strangers, this did nothing to alleviate the negative contagion of fatness.

Disgust toward obesity can be directed inward as well as outward, and when turned upon oneself becomes one of the most insidious causes of eating disorders. Eating disorders usually begin during adolescence and, though both men and women are affected, women far outweigh men in numbers afflicted (in North America women outnumber men ten to one). Although it is commonly assumed that eating disorders are the scourge of white, middle-class teenagers, an increasing number of women from various backgrounds aged fifty or older are turning up at treatment clinics in the US.[28]

The two most common eating disorders are bulimia and anorexia nervosa. Bulimia involves episodes of out-of-control eating—bingeing—typically on high-fat, high-carbohydrate foods (whole boxes of cookies, containers of ice cream, and trays of pizza may be devoured in fifteen minutes or less), followed by feelings of guilt, remorse, and self-loathing with a strong desire to undo the damage, usually in the form of induced vomiting or laxative abuse—"purging." These episodes can take place from several times a day to several times a week.[29] The incidence of bulimia is highly influenced by media and peer pressure for thinness, which is why it is especially prevalent in North America and Europe and has historically been almost nonexistent in Polynesian, Arabic, and African cultures, where thinness is not prized—though this is changing as Western ideals are imported. Despite their intense binge-and-purge behavior, bulimics are usually of normal weight. No matter: bulimics feel they are fat, and are typically depressed and disgusted by their bodies and by their actions.[30]

Unfortunately, a negative self-image only fuels the binge-and-purge behavior. Psychiatric evaluations have shown that the bulimic's self-disgust makes her want to punish herself by both perversely trying to make herself fat (bingeing) and by abusing herself physically (purging).[31] Anorexia nervosa, which drives people to become skeletally thin and refuse to eat almost entirely, is also fraught with feelings of body-loathing and self-disgust. Several studies have shown that people with eating disorders are more easily and intensely disgusted than healthy eaters, especially by triggers that involve food or bodily functions and fluids. Moreover, the more serious a person's eating disorder is, the more sensitive he or she is to disgust.[32] Bulimics and anorexics are also especially disgusted by seeing obese people. However, being highly sensitive to disgust is not the cause of eating disorders.

The first and only experiment to test whether disgust sensitivity was causally related to eating pathology showed that when women with eating disorders were covertly exposed to the aroma of "four nasty-smelling Limburger cheeses" they, just like a control group of healthy eaters, showed increases in disgust.[33] But the cheese manipulation had no effect on their body self-image, eating behavior, strategies to change their weight, or reactions to high-calorie foods. In other words, disgust needs to come from within and to be aimed at oneself for it to play a role in the dark side of eating.

Why should obesity be considered disgusting, and what facet of disgust does it spark off? One might assume that images of animalistic eating or abnormal body proportions are the central triggers of our revulsion toward the obese, but this has not been experimentally verified. Rather, several studies have shown that our repugnance toward obesity is implicitly associated with our aversion to disease.

In an experiment conducted at the University of British Columbia, an Implicit Association Test was used to investigate people's attitudes about body weight and disease.[34] The test has been used in hundreds of experiments to tap into true feelings on everything from racism to religion. Using a computer and specific keyboard presses, the test asks you to make "good" versus "bad" judgments associated with various concepts. First, you practice with some non-controversial topic, such as associating flowers and sunshine with "good" and death and disgust with "bad." Then the task changes to a controversial subject—fat people versus thin people, for instance—but the screen presentation and computer keys associated with the concepts also change. This forces you to pay more attention to the task and in so doing measures your true feelings, because you will take longer to press the "correct" key when you have to think about your answer in order to make the socially appropriate choice, than you would for your honest and automatic attitudes. Therefore, if you are slower to connect "good" with fat than you are "good" with thin, your true feeling is that thin people are better than fat people. If you dare, you can find out what your real feelings are on a variety of controversial topics at Project Implicit online, https://implicit.harvard.edu/implicit/.[35]

The experiment with University of British Columbia undergraduates linked the concepts "fat" with "disease" and "thin" with "healthy," versus the opposite (fat with healthy, thin with disease), and showed that participants had a stronger association to the concepts of "fat" and "disease" than "thin" and "disease." What's more, this association was especially strong if the participants had first seen a slideshow depicting germs, infections, and the prevalence of illness.[36] A follow-up study further showed that the more someone was worried about getting sick, the more likely they were to be disgusted by pictures of fat people.[37] That is, fat people are aversive because they insidiously conjure thoughts of germs and sickness. Fat is also usually perceived as ugly, and ugliness can disgust us for disease-related reasons as well.

THE GOOD, THE BAD, AND THE UGLY

Being viewed as ugly is often the result of having asymmetrical features—one eye bigger than the other, a crooked nose, uneven shoulder height. Lopsidedness can indicate that one's genetic makeup is faulty or that one's immune system is frail. If your immune system is weak, stresses during your gestation may have a negative effect on your physical appearance. For example, the energy your developing body had to expend to fight off a virus could have caused faulty formation of your eyelids if you didn't have enough resources in reserve to develop properly. Unattractive birthmarks might also mean that you harbor an unhealthy recessive trait. Therefore, ugliness is rejected because physical unattractiveness is often correlated with ill-health, particularly at a genetic level.

It has been proposed that our repulsion toward various people—immigrants, the obese, the ugly—is a by-product of our motivation to recoil from disease and that as such it is biologically adaptive. This bigotry is propelled by what has been dubbed our "behavioral immune system."[38] The behavioral immune system is a psychological mechanism that some scientists have proposed evolved to motivate us to withdraw from people who look like they might spread disease or pose a reproductive fitness risk. Indeed, several studies have found evidence for a link between illness, disgust, and social rejection. People who believe they are at an increased risk of getting sick, either because they are hypochondriacs or because they've just seen pictures of germs, react more negatively to foreigners and fat people. They are also less likely to be friends with people who have physical disabilities, and merely being exposed to pictures or information about illness leads them to be less agreeable, less sociable, and to automatically engage in more avoidant arm gestures, such as jerking back.[39] Furthermore, as you read in the last chapter, people who are vulnerable to disease, such as women in the early stages of pregnancy, show an intensified bias against foreigners.[40]

There is no question that it is adaptive to avoid that which is diseased, contaminated, or harboring illness, and many things which disgust us obviously signal a health risk—a beggar covered in red sores, the smell of sickness on someone's breath, the sounds of a phlegmy cough. All of these are valid signals of disease, and avoiding people who evince them helps protect our well-being. But is recoiling from someone dressed in unusual garb, speaking a language we don't understand, or who is fat or has a weak chin an instinctive psychological signal to preserve our health? And does it justify social antipathy?

My opinion is that although we act in antisocial ways that can be explained as protective or "preventative" against disease, I do not believe that the "behavioral immune system" is an evolutionarily evolved human trait motivated by prophylaxis against sickness, nor do I think it explains why we are prejudiced or behave cruelly toward others. For one, there is copious evidence that, just as disgust is not instinctive, we do not have an instinctive fear for our health. If a trait is hardwired, young children will do it, but children do not innately avoid signs of sickness and death—rather, as discussed in chapter 2, they revel in muck and mucus and are eager to approach "dead kitties." Our responses to illness and disease-causing agents have to be learned. In fact, disgust as a disease avoidance mechanism appears to fail us when we naturally need it most.

Early childhood and old age are the two periods in life when we are most biologically vulnerable to infection and therefore *should* be most susceptible to feeling disgust toward anything from frogs to foreigners, yet these are the two periods in life when we are least disgusted. It is also the case that though distorted features or the indication of low levels of various hormones—a weak chin can suggest low testosterone—may indicate a less than perfect mate, many people with deformities are clearly the victim of defects which have nothing to do with genetic hardiness. Indeed, many examples of bodily mutilation or distortion are self-inflicted, such as the current fashion for earlobe gauging. Ugliness is also culturally relative. The "beauty" of Cleopatra has been disputed by historians from the time of Plutarch to today, with one of the arguments being whether her nose was too big and therefore ugly, or not big and patrician enough and therefore ugly.[41] And when it comes to decisions about whom we'd like to have sex with, a recent study found that we often behave in risky ways when we make these choices. If people are reminded that they will eventually die, they are more likely to say they will have casual sex with a stranger and unprotected sex at that.[42] This is counterintuitive to the existence of a behavioral immune system; if such a mechanism were in place, reminders of death should make people more wary of strangers, especially having reckless sex with them, since sex is one of the most common routes of disease transmission.

There is no disease-sensible reason to be disgusted by obesity either. You can't get fat by standing next to or touching someone fat, and fatness is not sickness, nor is it true that fat people are dirty or unclean. Furthermore, apart from a heightened risk of heart attacks, strokes, and diabetes, which are not contagious, the health of the obese is no different from the health of the thin. Overeating and a sedentary lifestyle can make anyone fat, no matter how ideal their genetics are. The "disgustingness" of fat is also culturally variable. Heavy bodies have been considered more attractive than thin bodies in many cultures, both historically and at present. The paintings of Peter Paul Rubens are indicative of how plus-size physiques were appreciated in seventeenth-century Belgium. In the United States today, African Americans will select a body silhouette as "ideal" that white Americans label "fat."[43]

When it comes to foreigners, although historically most of the world's population lived in relatively isolated, fixed communities, many of us now live in multicultural urban areas where people of other ethnicities whom we encounter live next door to us and therefore share most of our germs and are as healthy as we are.[44] There are also many advantages to be gained by engaging with dissimilar people. Interchange with outsiders can promote new technologies, genetic diversity—which is much healthier than inbreeding—new alliances, the interchange of ideas, and increases to the economy. Therefore, avoiding diversity is limiting to our potential. Finally, no matter the basis, justifying antisocial and discriminatory behavior on biological grounds has grave social implications and consequences, as the history of eugenics and sterilization of the mentally handicapped attests.

Scientists who support the idea of an innate psychological defense mechanism against illness accept its problems, and defend the overgeneralizing errors the behavioral immune system makes on the grounds that it is better to be predisposed to bigotry than to be blithely indiscriminate since, evolutionarily speaking, the perils of being unprejudiced outweigh the benefits of impartiality. You can die if you catch the foreigner's disease or be a genetic dead end if you mate with a partner who carries a lethal trait, but being unbiased gives you no reliable evolutionary advantage. Proponents of the behavioral immune system also justify the social mistakes it generates with the argument that this mechanism evolved when the foreign and ugly would have been accurate threats of disease, and that not enough evolutionary time has elapsed for these reactions to be weeded out of the gene pool, nor might they ever be as prejudice doesn't appear to incur any biological disadvantage. In other words, our erroneous repulsion toward the fat, foreign, and ugly is akin to other disgust mistakes, such as being repelled by spiders—not particularly useful now, but not detrimental and evolved from a time when it might have been beneficial. Nevertheless, in my opinion, the idea that xenophobia, racism, and prejudice are biologically advantageous because the foreign, fat, and ugly might infect us or wreak havoc on our genetic fitness doesn't work. Rather, I believe the root of our repulsion toward these "misfits" comes from a far deeper concern.

STRANGE AND CHANGE

What makes foreigners, fat people, and Ugly Bettys notable is that they are strange—and it *is* adaptive to be leery of the unknown. "Strange" means that something is unusual or unknown, and the unknown may be hazardous, whether it is food, culture, geography, new electrical appliances, people, or bacteria. In this regard, given the rise of obesity in North America, fat may not be strange for much longer, and the extremely thin may soon be perceived as more disgusting than the obese. Indeed, it was recently shown that women who are healthy eaters (don't have eating disorders) rate very thin body figures as more disgusting than obese bodies.[45] In any event, it is wisest to cautiously explore what is unfamiliar or wait for the new to show itself rather than dashing up to it zealously or ignoring it. This wariness is not justification from biology to condemn foreigners or exclude the fat kid, merely an adaptive reason to be careful in the face of the strange, no matter what it is. However, it is more beneficial if this caution is tempered with enthusiasm rather than avoidance.

Today we live in a world with greater global interactivity than ever before. We virtually connect, fall in love, and do business with people all over the planet. Steven Pinker, the renowned psychologist and author of *The Better Angels of Our Nature: Why Violence Has Declined*, discussed with me at a recent conference how, despite news that continuously announces war and bloodshed, there is less conflict and violence worldwide than there has ever been in human history. The world political climate has never been better, and our global interconnectedness is the reason why.[46] But interconnectedness also has a dark side. The people we are most likely to reject and condemn, both historically and currently, are not people from exotic lands but our close-enough-to-touch neighbors.

Civil wars, insurgencies, and genocides are committed by a culture toward another very closely connected culture—a culture that, from the perspective of the aggressor, differs only in its worldview, the framework of ideas and beliefs through which the world is interpreted and that gives life meaning and structure. Worldview threats are religious, political, or ideological; they are not due to hygiene, weight, birthmarks, or cuisine. Indeed, on his CNN show of April 25, 2004, "Horror and Hope: Rwanda, Ten Years Later," Anderson Cooper stated: "It wasn't so much the speed and savagery of the massacres that shocked the world, but the fact that it happened in a nation where the two main ethnic groups, that shared so much in common and lived in relative peace for most of their long history, could turn on each other with such brutality." The condemned group does not pose a threat of making other people sick. It poses a threat to the "order," power, and belief structure of the dominant group.

Strange and change disrupt our standard frame of reference. The arrival of immigrants augurs that a change to what we are accustomed to is looming. We don't like change, never mind whether we will be disadvantaged by it ("they're taking our jobs"). In its simplest terms, it's the "who moved my cheese?" syndrome. We like the status quo and want to maintain the traditions we are used to. It isn't what they eat for dinner, but rather the unknown consequences of foreigners moving in next door that is really unnerving to us.

The obese, deformed, and ugly are strange and threatening because their bodies are distorted, and this distortion reminds us how easily our bodies can be mutilated. There is a peculiar disorder called acroto-mophilia, in which people are only sexually aroused by amputees, and another even more bizarre condition called apotemnophilia, where people wish they were amputees. If you have apotemnophilia you can find websites linking you to kindred spirits, with helpful tips on how to have an industrial accident if you can't convince your doctor to amputate your limbs for no reason. These desires are symptoms of psychiatric disorders. By contrast, most of us are repelled by the deformed, foreign, fat, ugly, and the mere thought of having our legs cut off, and this is because all these physical conditions subconsciously provoke our utmost fear.

THE GRIM REAPER

A student posed in a wheelchair rolled up to various passersby at the entrance to the main university library and asked them if they'd be willing to help her complete a survey for a class assignment. On another day, the same student walked up to various passersby and asked if they'd be willing to help her complete a survey for a class assignment (no wheelchair was in sight). The student approached eighty-five different passersby in total. How often do you think people agreed to help the student with her survey? And do you think their willingness had anything to do with whether she was in a wheelchair or not?

This was the setup in an experiment conducted at Bar-Ilan University outside of Tel Aviv.[47] The results showed that more passersby overall were willing to help the student in the wheelchair than the physically capable student who walked over to them. But the experiment had a catch. The passersby had first been nabbed by someone else who was secretly part of the experiment, and given a flyer. The flyer was from a fictional organization and either read, "Are you concerned about death? We can help! Call us and we can ease your suffering both physically and spiritually," or, "Are you dealing with back or muscle pain? We can help! Call us and we can ease your suffering both physically and spiritually." The surprising result was that passersby who got the flyer about death were much *less* likely to help the student if she was in a wheelchair than if she approached them on foot. And they were especially unhelpful compared to passersby who got the flyer about muscle pain. Only 58 percent of people who got the death flyer completed the survey for the student in the wheelchair, compared to 89 percent of people who got the flyer about muscle pain.

Other research and probably your own experience exposes the fact that healthy people are ambivalent toward the handicapped, though when social pressure prevails guilt tends to make us act sympathetically. Yet, in spite of the high social pressure of a large public meeting place, the death flyer dismantled social niceties and exacerbated heartlessness.

Why? The answer lies in what I believe is the raison d'être for the emotion of disgust.

We are the only creatures who experience emotional disgust, and we are also the only creature that knows it will die, though this realization does not crystallize until mid- to late childhood, which is why children don't fully experience disgust. The deformity of others—obesity, ugliness, being handicapped—reminds us that we inhabit fragile physical forms which can be distorted, mutilated, and—the ultimate terror—annihilated.

In 2001, psychologists from the University of Colorado, the University of Arizona, and Brooklyn College conducted several experiments to examine the idea that death and disgust are fundamentally linked. In one experiment, half the participants were asked to reflect on the emotions that the prospect of their own death aroused in them and to jot down what they thought would happen to them physically as they died. The other participants were asked to think about television.[48] All the participants were then assessed for their disgust sensitivity, and it was found that as a function of what they had been previously asked to think about, their disgust sensitivity changed. Thinking about one's own death induced higher disgust-sensitivity scores, especially disgust pertaining to maggots, cockroaches, and body fluids, than thinking about television did.

In another experiment by the same researchers, college students were told either to contemplate their own death or another aversive experience—severe dental pain—and then asked to read an essay titled "The Most Important Things I Have Learned about Human Nature," which had supposedly been written by an honors student at another local university.[49] For half of the participants, the contents of the essay emphasized the special potential and uniqueness of humans and the humanistic spirit, while the other participants read a Darwinian essay that stressed the similarity between humans and other animals. In other words, one essay reminded people about the inherent connection between humans and animals and the other stressed human uniqueness. All participants then gave their reactions to the essay and rated how likable, intelligent, and well-informed the author was, as well as how much they agreed with the author's opinion. The results were intriguing. Undergraduates had substantially more positive reactions to the "humans are unique" essay and its author compared to the Darwinian essay, but only if they had previously been asked to think about their own death. Students who had thought about dental pain were equally predisposed to both essays and their authors.

In a similar experiment, participants looked at either disgusting pictures, such as a man urinating or a person vomiting, or neutral pictures such as a book or a clock, following which they were given word fragments like DE_ _ or COFF_ _ and asked to complete the words.[50] The pictures that the participants had previously seen altered the way they completed the words. Those who had viewed disgusting photos were more likely to complete the words in relation to death—"DEAD" or "COFFIN"—whereas participants who had looked at clocks and books were more likely to fill in the blanks to spell out "DEAL" or "COFFEE." In other words, reminding people that we are animals makes thoughts of death rise to the surface, and feeling disgusted activates thoughts of death as well.

Disgust and fear of death also lie at the heart of in-group favoritism, prejudice against foreigners, and negative stereotyping. When nationalistic Canadian university students read an essay that denigrated Canadian ideals like hockey, socialized health care, and politeness, they were more likely to complete word fragments like SK_ _ _ as death words (e.g., "SKULL" versus "SKILL") and to react faster to seeing death words such as "grave" and "buried," compared to other negative words like "punish" and "fight," than they were after reading an essay belittling Australian values.[51] That is, thoughts of death specifically, and not just any negative emotion, rose to the surface when the student's Canadian worldview was threatened. Being reminded of our mortality makes us especially favorable toward people like us—the in-group—and especially unfriendly toward people who aren't like us—the outgroup—regardless of how superficially they are connected to us and even when it goes against our self-stated values. Thinking about one's own death has even been shown to make white, ostensibly liberal college students more sympathetic toward explicitly racist white employers, and to encourage the use of negative stereotypes (rude, arrogant) against students they had nothing to do with at a rival college.[52]

These findings have chilling implications. Campaigns that galvanize potential recruits by reminding them of threats to their mortality—"the terrorists will kill you"—will be especially likely to incite vitriol and violence. The Florida preacher Terry Jones, who ignited an international conflagration in 2011 by his much publicized burning of the Koran, attracted his followers by preying on growing uneasiness and anti-Islamic hostility surrounding the ten-year anniversary of the terrorist attacks of September 11, 2001.

At the opposite end of the religious rhetoric spectrum, the act of charity is often directly linked with salvation from death. For example, at the entrance to Jewish graveyards there is always a charity box inscribed with the biblical proclamation "Charity saves from death" (Proverbs 11:4). Most of us implicitly hold this feeling regardless of our religion or religiosity. Pedestrians stopped in front of a funeral parlor were found to be more generous to charitable organizations than those stopped on a street where there are no visible reminders of death.[53] But this charity is limited by self-protectionism. American college students who were primed with thoughts of their death gave larger monetary donations to charity than those who had been primed with thoughts of dental pain, but this generosity only extended toward an American aid organization. Reminders of mortality did nothing to increase generosity toward an identical *international* aid organization (donations were about half of what was given to the American charity).[54]

Real-life events pertaining to death and one's worldview give us another lens through which to witness how fear of our mortality intensifies out-group antipathy and in-group allegiance. After the terrorist attacks of 9/11, there was an unprecedented spike in American nationalism: patriotic songs filled the airwaves, stores couldn't keep up with the demand for American flags, and New Yorkers turned up for jury duty in record numbers. An investigation into American responses to 9/11 showed that how strongly someone believed in a just world—that people get what they deserve, and deserve what they get—was correlated with their anti-Islam and anti-Arab hostility. Americans who were more affiliated with the idea of a just world were most distressed by the attacks, most vengeful against the people and countries responsible for the attacks, and endorsed the most aggressive retaliation against "those responsible."[55]

Evidence that our brain responds differently to in-group versus out-group members as a function of reminders of death has also been shown. In a recent experiment by Jamie Arndt and his colleagues at the University of Missouri, white college students were asked to think about either their own death or dental pain and then were shown pictures of black and white male faces making either happy or angry facial expressions. The electrical activity of the student's brains was measured while they looked at each of the faces. The results showed that when white people had been primed with reminders of their mortality their brain activity indicated more intensive categorization of faces by race and a decrease in the extent to which white angry male faces were perceived as threatening.[56] In other words, after being reminded of their death, differentiating people as belonging to their in-group (white) or not belonging (black) became more salient, and threats from in-group members (angry white faces) were downplayed. When you are threatened with death, you keep to your own and assume more safety from those in your fold.

The connection between death, disgust, and our view of the world can be explained by a compelling idea called terror management theory, which was put forward in 1973 by the cultural anthropologist Ernest Becker in his Pulitzer Prize-winning book *The Denial of Death*,[57] and then further developed by the psychologists Sheldon Solomon, Jeff Greenberg, and Tom Pyszczynski.[58] In a nutshell, the theory explains that humans confront a horrendous existential challenge in the awareness of our mortality and the fact that we have not yet been able to solve the problem of death. One of my students poignantly confessed in class one day how she remembered the moment when she was five years old and first realized that she would die and how she "just cried and cried for hours."

The terrible and terrifying inevitability of our death leads us to develop various psychological strategies to cope with this truth, which include denying it, suppressing it, and generally avoiding thinking about death as much as possible. When reminders of our inexorable demise rear their ugly heads, we are driven to defend ourselves psychologically, and in bolstering our defenses we do many things, such as shunning and avoiding groups and people who remind us of our mortality, shoring up our views of an ordered and stable universe, favoring people who agree with our worldview, and rejecting—even destroying—those who don't. However, unlike the concept of a behavioral immune system, terror management theory does not provide excuses for the terrible things we do to one another. Rather, Becker blamed most of the evil that humans perpetrate upon one another as due to our need to deny death. Terror management theory simply offers an explanation for why humans do such atrocious things, and why we are so threatened and repulsed by certain groups and individuals.

Our disgust toward people and stimuli that remind us of our mortality is based on a contention that has been put forward by a number of longtime disgust researchers—which I have been strongly persuaded by—that disgust is fundamentally about our awareness of our own death and our terror of it. The emotion of disgust arose from our need to protect ourselves from triggers that remind us of this truth—such as our animalistic nature—and put us in its harmful way—such as disease. Disease is a primary motivator of disgust, but it is not the psychological construct that controls it; our fear of death is. In fact, when Jonathan Haidt and his colleagues were devising the Disgust Scale presented in chapter 2, they found that the level of disgust experienced in response to items involving contact with dead bodies and a person's fear of their own death were the strongest predictors of someone's overall disgust sensitivity.[59] That is, the higher your score on the Disgust Scale, the more you fear your own death. Furthermore, when research participants were reminded of their own death, as in the previous experiments, their scores on the Disgust Scale rose compared to if they had been thinking about something else.[60]

The emotion of disgust emerged in humans to protect us from the problem of our death both physically and psychologically, and this is why it motivates us to recoil from scabby sores, piggish eating, and people who threaten our way of life. We treat other people despicably when we are disgusted by them, and we are disgusted by them when they stir thoughts of our inevitable demise, or threaten the social constructs and illusions that keep the truth of our mortality at bay. Disgust has been called an emotional protest against death.[61] Disgust says, "I reject this," or "I reject you," and in so doing strives to defend us from the potential path of extermination that "this" is portending, either symbolically or in reality. Rejection is the fundamental stance behind all that is disgusting.

Given our fundamental terror of and repulsion toward death, one would expect that we would do everything possible to avoid disgusting experiences and reminders of our mortality. Yet, paradoxical and quirky creatures that we are, as you will now see, we are instead perversely attracted to and lured by death, destruction, and that which disgusts us most.

Chapter 6

HORROR SHOW

The sun is rising in a red sky. An imam is chanting a prayer. Sheep are being herded and workers in turbans are digging at an archeological site in northern Iraq. Suddenly a boy in a red keffiyeh is running to the lead archeologist, gesticulating. Something has been found. The archeologist hurries to the location. Lamps, arrowheads, coins are dusted off from a tray, and then a strange amulet. The archeologist digs further into the opening at the base of a mound and extracts a small, sand-encrusted rock with something attached to it. He breaks off the piece and dusts it—and the black sculpted face of the devil is revealed.

This is the opening scene of *The Exorcist* (1973). It is rated by film critics as one of the top horror movies of all time, and it is also one of the top money-making films ever made. Ranking number nine, *The Exorcist* comes in just behind *Jaws* (1975) as the highest-grossing horror movie.[1]

When *The Exorcist* hit the silver screen, the audience screamed, some fainted, some vomited, and one man reportedly broke his jaw on the seat in front of him. Vomiting while watching *The Exorcist* is an excellent example of the vomit empathy response, as Linda Blair repeatedly regaled the camera with her devilish green spew. Fainting demonstrates the drop in blood pressure that occurs with disgust, and breaking one's jaw may be the unfortunate consequence of an overly exuberant "ahh-ugh" reaction.

Not only can you hurt yourself by flailing about in disgust, but horror movies may also be bad for your health. In an experiment conducted in Britain, one group of participants was assigned to watch the grisly 1974 hit *The Texas Chainsaw Massacre* while a control group of volunteers sat in a room with banal reading material for the same length of time (eighty-three minutes).[2] Blood samples and other cardiovascular measures were taken from both groups, before and after their session. As expected, the heart rate and blood pressure among those who had watched *The Texas Chainsaw Massacre* increased by about 20 percent, whereas in the control group there was no change. The bigger issue, however, was that blood samples taken from those who had watched the film showed markedly elevated levels of leukocytes, the white blood cells our body releases to fight off invading pathogens. Blood samples from the control group were normal. This means that watching *The Texas Chainsaw Massacre* set off the body's immune response when there were in fact no pathogens present. The problem is that this false-alarm immune reaction and release of leukocytes produces a temporary depletion of leukocytes which would have been available in the case of a real health threat. In other words, if you watch a horror movie during flu season and then pick your kids up from day care, you'll be more likely to catch the bug going around than if you had watched *Caddyshack* (1980). Where does the demand for watching movies that are clearly aversive, and may even be unhealthy, come from?

A LITTLE HORROR HISTORY

The first horror motion picture ever made premiered in Paris in 1896. It was a two-minute short called *Le Manoir du Diable* ("The Haunted Castle"). Audiences would then have to wait another twenty-four years for the first full-length horror feature. In 1920, the terrifying German expressionist film *The Cabinet of Dr. Caligari* was released. A few years later came *The Hunchback of Notre Dame* (1923), starring Lon Chaney as Quasimodo, followed by another Lon Chaney hit, the ever-playing *The Phantom of the Opera* (1925). By the 1930s, horror had taken a firm hold on haunting the night, and masterworks such as *Dracula* (1931), *Frankenstein* (1931), *The Mummy* (1932), *The Black Cat* (1934), and *The Werewolf of London* (1935) were shown to screaming audiences in North America and Europe. But the horror film industry has not escalated linearly in viewership or production. Rather, sociohistorical events, especially war, have driven or deadened the genre's popularity.

In the early 1940s, *The Wolf Man* (1941), starring Lon Chaney Jr., Claude Rains, and Bela Lugosi, as well as many other horror films were big box-office hits, but only in countries like the US, where the fighting of World War II was out of plain sight. In Europe, where death and destruction were everywhere, the nightmare of reality took the interest out of paying to watch fictionalized horror. In North America, the only direct images the public saw of war were of heroic soldiers mowing down Nazis, but the knowledge that real slaughter was occurring and the constant expectation of news that a loved one had died lurked in people's minds. What horror movies did was provide a fiction of control in which the terror could be captured and vanquished. Therefore, when the war—and this anxiety—ended, so died the horror movie industry. According to Curt Siodmak, a screenwriter and novelist whose credits include *The Wolf Man*: "The day the war ended, the bottom of the horror movie industry fell out . . . Horror pictures couldn't even be given away."[3]

The horror movie industry was revived by Cold War anxieties in the 1950s and 1960s, as the abstract and protracted dread of an unseen enemy and annihilation by nuclear holocaust loomed. In *Invasion of the Body Snatchers* (1956), giant pods from outer space symbolized American fears of the "Red menace" and fears about the deforming effects of nuclear weapons. An interesting exception to this cycle occurred during the Vietnam War. For the first time, the American public was inundated with real horror as television news brought scenes of dead and dying American soldiers and bloodstained battlefields into the living room nightly. This time, instead of spiking when the combat was across the globe, horror movie interest declined and a new brand of horror film—as sociopolitical commentary—emerged.[4] The zombie cult classic *Night of the Living Dead* (1968) is well-known to be a critique of the mayhem and loss of humanity that occurred during the Vietnam War.

In the mid 1970s, horror movie production began to increase again, but this time with a new twist. Rather than fright being the central draw, another emotion was competing for star billing: disgust. Early horror films dwelled primarily on eliciting fear, with minimal portrayal of graphic murder, torture, and gore. In *Psycho* (1960), you never directly see Norman Bates stab Marion Crane. But in horror films produced after 1975, for the first time gore was on equal or greater display as terror, and nowadays, with terrorism as the skulking menace, horror movies that splatter the screen with as much viscera and mutilation as possible are more popular than ever. In fact, a new subgenre called "extreme horror" has emerged, with so-called Splat-Pack films such as the *Saw* franchise (2004–10), *The Hills Have Eyes* (2006), and the *Hostel* series (2005–11). These films are rife with no-holds-barred carnage, torture, and protracted bloodshed. The director of the 2007 remake of *Halloween*, Rob Zombie—this is his real name—explained that in his films "violence isn't gratuitous—it's the point."[5]

The limits on cinematic disgust seem to be bounded only by the imagination. Nevertheless, there is something special about horror movie mutilation that has an appeal where non-horror gore does not. People will quickly turn off a slaughterhouse documentary that has the blood-and-guts content equivalent to a popular horror film.[6] The difference is the reality factor. Horror movies give explicit signals of their fiction—eerie music, bizarre camera angles, and fantastical plots—which allows the viewer to indulge in the horrific. Fiction enables disgust to become entertainment. Horror mockumentary movies such as *The Blair Witch Project* (1999) unintentionally made this point when dedicated horror fans complained what a letdown this "reality horror" film was.

When John Carpenter's first "slasher" movie about a murderous supernatural psychopath on a killing rampage—*Halloween*—was released in 1978, it immediately became a huge box-office hit.[7] The film, which was made on a budget of only $320,000, grossed over $60 million worldwide that year. And profit drives creation. In 1980, sixty-six horror movies were released worldwide.[8] In 1990, 229 horror movies were made; by 2000 the number had increased to a modest 358; but since 2000 the production spree has escalated at a startling rate. In 2006 no fewer than 874 fear flicks were released, and by 2010 the number had risen to 1,017 (excluding TV miniseries and rereleases).[9]

The reason for the skyrocketing production of horror films in recent years, especially those that feature extreme gore, isn't entirely clear. One explanation is that, as during World War II, we are visually shielded from the atrocities that are taking place elsewhere on the globe. Vietnam video journalism was an exception. Indeed, the only shocking imagery most of us have seen from the current Iraq and Afghanistan wars are the Abu Ghraib photos of 2006. Another factor is that demand drives production, and serious horror movie fans claim to become "addicted" to watching them. Their claims are not without basis. The high-intensity emotions that are ignited by horror (terror and revulsion) give viewers an adrenaline rush and release of endorphins (the body's natural opiates) which, though much weaker than a speedball (injecting a mixture of heroin and cocaine), are in the same physiological ballpark and apparently as enjoyable to some people.[10] Addiction means that the high is followed by withdrawal, which can only be sated by more of the drug—horror films. This cycle of need intensifies with each repeating loop. And since watching the same movie over and over is not nearly as rewarding as getting a fresh buzz, there may be no limit to how many new horror movies will be made as long as an "addicted" audience keeps craving more.

THE LURE OF HORROR

Many theories have been proposed for why we are drawn to such a seemingly counterintuitive and even counter-health activity as deliberately revolting and terrifying ourselves. Evolution-based theories argue that we do things for a "good" reason. That is, some adaptive advantage for our survival and reproductive fitness must be conferred if a given behavior is maintained. But if watching horror can make you more susceptible to illness, how could it be biologically favorable? More to the point, given our abhorrence of death and disgust, what motivates us to deliberately want to watch such terrifying and repulsive cinema?

One popular explanation for the lure of horror is that we get relief from the stresses and anxieties of real life by viewing it. As during World War II, news of war across the globe and today's constant furtive menace of terrorism produces a culture where fear is ever-present, and therefore a release from fear is necessary if we are to cope with daily life. Theorists from disciplines ranging from media studies to psychology propose that horror movies provide an outlet for our fears and offer temporary relief from our worries that lurk in the shadows. As Alfred Hitchcock said, "I aim to provide the public with beneficial shocks."[11] The reason we come back for more is because the catharsis works, but because the relief, as well as the high, is only temporary we'll need another fix again soon.

In addition to geopolitical apprehension, personal anxiety is another motivation to expose ourselves to the gruesome and terrifying. That is, horror movies can offer catharsis from our personal traumas. The first time in life when many "monsters" and anxiety torment nearly all of us is during the teenage years, so it is no accident that schools and school-age characters dominate contemporary horror movies, on the screen and in the audience. *Carrie* (1976), Brian de Palma's masterful film of Stephen King's 1974 novel, tells the story of an outcast teenager who is the subject of intense derision and cruelty from her classmates. Carrie then finds herself equipped with supernatural, telekinetic powers and inflicts murderous revenge at her high-school prom. An audience of teens can identify with the characters and the environment. School as a backdrop for terror and horror provides a fantastical extension of daily anxieties about failure, social exclusion, and lack of control. Watching horror and violence taking place in a school setting not only gives adolescents relief from their own routine horrors, it gives them an outlet for their feelings of hostility toward peers, teachers, and parents.

HORROR THERAPY

The psychological benefit of horror movies for teens is similar to the respite that young children get from the macabre fairy tales that make for *comforting* bedtime stories. Despite the tender age of their target audience, fairy tales are anything but tame, with monstrous wolves devouring their victims, as in "Little Red Riding Hood" and "The Three Little Pigs," and witches burning children alive, as in "Hansel and Gretel." Fairy tales are rife with terror and gruesomeness. But, importantly, fairy tales portray unempowered and vulnerable characters who, through their resourcefulness and ingenuity, overpower the evil forces. The stories conclude with the protagonists' victory and safety.

Fairy tales offer psychological succor by enabling the child to associate with the stories' heroes so that she is now armed with the courage to neutralize the terrors that lurk beneath the bed and in her closet. In the same way, slasher films typically depict a hero or heroine who is initially the most timid or helpless among the bunch, but who then emerges as the most valiant and vanquishes the "monster." Moreover, in many slasher films, the parents of the teen protagonists are portrayed as useless and out of touch ("parents just don't understand . . ."), playing to the classic teen–adult struggle. Slasher films, with their teenage actors, school-based backdrops, and parental ineptitude, provide a fantasy setting where adolescents can be bolstered with feelings of control and power over their worries, which helps them to gain mastery over the tumult in this difficult phase of development. Whether watching horror movies actually decreases teenage depression, anxiety, the number of hallway brawls, or dinner table battles isn't known. But it has been shown that when teens act violently, horror films can be used therapeutically.

In 1990, a thirteen-year-old boy who was committed to a psychiatric facility after drinking through the liquor cabinet and then destroying his aunt and uncle's home with an axe was healed by horror. During his therapy sessions, the boy confessed his resentment of his uncle (his legal guardian) and repeatedly mentioned scenes from the horror films he was "addicted" to. The boy was an especially ardent fan of the psychopathic murderer Freddy Krueger from the *Nightmare on Elm Street* series, and so in an unprecedented move the psychiatrist decided to use *A Nightmare on Elm Street 4: The Dream Master* (1988) in psychotherapy with him. The psychiatrist watched portions of the film together with the boy, helping him to come to the realization that Freddy's violence and hatred stemmed from his having lost his mother at a young age. The psychiatrist then pointed out that the boy might understand how Freddy felt, since he had been placed in his aunt and uncle's home after being abandoned by his own mother at the age of nine. Continuing treatment in this way, the boy and his psychiatrist discovered and worked through many of his emotional problems, and along with family therapy the boy was successfully treated.[12]

Identifying with horror movie characters, however, doesn't always end well, as was the case when a fifteen-year-old boy killed a seven-year-old in order to use his fat to create "a flying potion." The teenager had apparently gotten the idea from the movie *Warlock* (1989), in which the villain boils the fat of a child to create an elixir that can make him fly. The fifteen-year-old had no previous criminal history, but when clinically examined he was found to be delusional and believed he was in the presence of the devil.[13] Psychotic disorders, where there is a true disconnect from reality, are very serious but also rare, and murderous rampages by avid horror movie watchers are therefore highly unlikely.

However, some personality types known as "gore watchers" may be unhealthily captivated by the carnage on the screen. Gore watchers get pleasure from horror movies specifically as a function of how much blood, guts, torture, and mayhem are depicted. Gore watchers are also most likely to fantasy-imitate their favorite character—the villain, to have negative attitudes toward women, and to display some psychopathic personality traits, such as low levels of empathy.[14] By contrast, "thrill watchers," who derive delight from jolts and suspense, get their enjoyment because they identify and empathize with the spills and thrills of the victims, and are generally of no danger to anyone. Parents of teenagers who seem to get extra pleasure out of horror movies may want to pay attention to whether it is gore or thrills that their child is especially titillated by. If it is the former, some monitoring or intervention may be worthwhile.

Theories that relief from anxiety and the catharsis that comes from watching characters act out their angst do well at explaining the attraction of horror, especially for adolescents. The teen years brim with struggles over impulse control, managing aggression, dealing with anxiety, hostility in school life, parental expectations, and of course burgeoning sexual urges. Indeed, sex is another major motivator for the lure of horror, particularly for teenagers.

LOVE AND DEATH

The cliché that boys take girls to fright flicks because it makes the girls leap into their laps so that the boys can comfort them, in the hopes that it leads to more contact sport, is partly true. But the reason why horror movies can lend a hand in budding romances is due to more than the opportunity for a little grabbing in the dark. Horror movies provide a means to demonstrate an important dimension of sex appeal, especially for teenagers and young adults—mastery of gender-role identity. In heterosexual hunting, how macho or feminine you are has a big impact on how attractive you are seen to be by your peers. And the way that young men and women act while they watch a horror movie together is an ideal opportunity for revealing and testing each other for their gender-identity mastery. Not only will their behavior influence how much they'll like each other when the lights go up, it will also influence how much they each like the movie.

Researchers at Indiana University tested first-year college students for how much they liked a particular horror movie, as well as an opposite-sex partner, as a function of how that partner acted when they watched the film together. The film was a 7.5-minute segment of *Nightmares* (1983) that was highly suspenseful, but not gory. In each opposite-sex duo, one member of the pair was a confederate—who played a prearranged role for the experiment—and the other was the unsuspecting participant. Depending on the condition the participant was assigned to, the confederate was instructed to act fearless, distressed, or indifferent to the horror on the screen. A "fearless" confederate sat casually slouched and shouted encouragements like "That's the idea . . . use the knife!" Acting "distressed" entailed fidgeting, nervous body scratching, and muttering "Oh my God." "Indifferent" confederates were quiet and expressionless. After watching the movie segment, the participants rated both their enjoyment of the movie and the sex appeal of their viewing partner.

It is a testament to the power of sex roles that how much the movie was liked was directly affected by how macho or feminine the male or female confederate acted. Men rated the movie as most enjoyable when the woman they were with acted distressed and as least enjoyable in the company of a fearless woman. By contrast, women enjoyed the horror movie least in the company of a squeamish man, and most in the company of an intrepid companion. Interestingly, when it came to how much the partner was sought after, natural good looks were able to take the edge off "incorrect" behavior. When the partner was judged as very good-looking, they were still considered desirable whether they yelped, sat stonily, or encouraged the film victims, but when the partner was not so beautiful, how they acted had a big impact on their attractiveness. In particular, when a so-so-looking guy acted tough in the face of terror, he gained lots of points in both positive character traits and how much his female companion wanted to see him again.[15] Playing macho while watching horror movies boosts the allure of men who aren't graced with irresistible looks. Therefore, if you don't look like Justin Bieber, you can still end up with star appeal if you take your date to a horror flick and show off your bravado.

The reason horror movies are such an ideal venue for demonstrating gender identity is because today's world offers so few occasions for adolescents to display this prowess. Most of us no longer spend our days hunting for food, fighting our enemies, or taking care of family, and therefore teenagers need to find other outlets to show they can do this if necessary. The teen girl needs to demonstrate her capacity for nurturance and compliance and the adolescent boy to demonstrate his resilience and bravery—and the safe venue of the movie theater or living-room couch provides the perfect setting.[16]

Besides providing an opportunity to display sex-role mastery, horror films also feature a lot of sex, and sex is sexy. In a scene analysis of horror movies, although violence occurred most often, nudity, kissing, fondling, and implied intercourse occurred more than nine times per film and "sex before death" occurred an average of three times per film.[17] In an investigation of how sex influences horror film enjoyment, men reported that sexual content made horror more enjoyable, while women stated that the sexual component didn't make a difference. This may not sound surprising, since men enjoy pornography more than women do,[18] but it was also found that sex makes horror scarier for men. Men rated horror movie clips with sex in them as more frightening than similar horror movie clips without sex. For women, sex on the screen had no impact on the scariness of the movie.[19]

Sex may not enhance horror movie enjoyment for women, but fear can make women sexually aroused. When women were shown frightening photographs—people pointing guns, terrorists, snakes poised for attack—and then watched a pornographic film, their sexual arousal increased above and beyond porn without frightening foreplay. However, when women viewed disgusting pictures, such as corpses and vomit, and then watched pornography, their sexual arousal decreased.[20] So, if a man wants to get his date "in the mood," watching a scary movie together is a good idea, but only if it isn't gruesome.

There is another blend of sex and horror that is especially alluring to teenage girls. I happened to be at the cinema on the same night as the premiere of one of the *Twilight* saga films, and my curiosity was piqued by the outrageously long line of teenage girls waiting to buy tickets. Why all the girls? I decided to find out what all the fuss was about and rented *New Moon* (2009). Teenage boys will likely be bored by this movie, but it may be worth it for them to sit through it for the amorous consequences that may ensue. The film tells the classic story of star-crossed lovers—a beautiful human girl and a beautiful vampire boy, complicated by a love triangle with a gorgeous Native American boy who turns into a werewolf. The vampires twirl and fling themselves with supernatural might and the werewolves lunge and gallop with mythical speed, but you see only slightly more blood in this movie than you would watching someone get a paper cut—which in fact happens and causes a vampiric frenzy. Basically the film is a hopelessly romantic teenage love story with death mixed in. And in this context death couldn't be more opposite from disgust.

In *New Moon*, as in so many romantic poems and stories, death symbolizes the devastation of heartbreak, the idealization of love's immortality, the yearning to give everything, including your life, to the object of your desire, and literally dying to be with the one you love. In the *Twilight* chronicles, death and monsters are not scary and grotesque but beautiful and romantic ideals—the apogees of erotic and passionate love. Watching this stuff with a teenage girl is excellent foreplay.

Cathartic relief, gender identity displays, and opportunities for fondling and sparking arousal all play a part in explaining why people, especially teenagers, are drawn to horror. But there is another reason, and I think the best one, for why horror is so appealing to the young and not-so-young alike.

THE THRILL OF IT ALL

> The devastating cyclone of events was finally over. Relieved and saddened, Sue stared at the blackened ground where the frightful house once stood. In the quiet, crisp, autumn air, she bent down contemplatively to place a small bouquet of hand-picked wild-flowers at the site of her former friend's dwelling, when suddenly a bloodied arm thrust through the ground, grabbed Sue by the wrist, and tugged her down to the earth below.

This is the final scene in *Carrie*, and the feeling you get when in that somber scene a grisly arm suddenly lunges for Sue is a rush of fear that can be enjoyable for the sheer excitement it produces. This is the same motivator that makes some people love to ride roller coasters, bungee jump, and sky dive. They know with a reasonable degree of certainty that their speed ride will not end in death, just as the horror movie audience knows that their theater experience will not lead to their imminent demise (unless they're watching *The Ring* (2002), where seeing it makes you die). These horrifying situations provide intense fear in a relatively safe context, leukocyte reactivity notwithstanding, and therefore are pure thrill. However, how much you want to be shocked and stimulated depends on your personality, and specifically a personality trait that is intertwined with your biology and changes as you age.

"Sensation seeking" refers to how physiologically aroused you like to be. People can be categorized on this dimension by answering "true" or "false" about themselves on statements like "I sometimes do 'crazy' things just for fun" and "I enjoy getting into new situations where you can't predict how things will turn out," as well as how much they're willing to take risks for the sake of such experiences, such as "I like to explore a strange city or section of town by myself, even if it means getting lost." If you answered "true" to all these scenarios, you are probably a high sensation seeker. High sensation seekers want novelty, complexity, and excitement, and they also enjoy negative emotions such as fear because of its intensity and accessibility. Fright-inducing exploits like riding roller coasters, bungee jumping, and going to horror movies are easy to do and easy to come by. Low sensation seekers would answer "false" to all the statements above, and they explicitly do not want or like excitement, novelty, risk, and intensity, nor do they enjoy negative experiences or fear. The majority of people are somewhere in between these two extremes.

How much you like to be stimulated is related to your baseline level of neurological arousal—meaning, how awake and active your brain naturally is. People who are naturally hyperalert don't like high-intensity stimulation because it can easily push them over the edge and become unpleasant, whereas people who are naturally set on low want more jazzing up. That is, high sensation seekers are actually functioning at a comparatively low baseline level of neural arousal and therefore need intense experiences in order to feel more alive, whereas low sensation seekers are internally revved up and therefore seek to avoid further or diminish current arousal. High and low sensation seekers, respectively, behave and modulate their environment in order to heighten or dampen their internal states so that they can enjoy life. High sensation seekers are more likely to become firefighters or race car drivers, and low sensation seekers to become librarians and gardeners. Not surprisingly, high sensation seekers watch a lot more horror and are more entertained by horror movies than low sensation seekers. And this is because it not only perks them up—they get more bang for their buck from it too.

In a recent experiment conducted at Friedrich Schiller University in Germany, brain activity was examined while participants who varied in their sensation seeking watched frightening and neutral scenes from movies such as *The Shining* (1980). Consistent with previous evidence that high sensation seekers operate at lower than normal levels of neural arousal and low sensation seekers operate at high baseline levels, the higher the participants' sensation-seeking drive was, the lower their neural activation was in visual areas and the anterior insula (where disgust is processed) while watching neutral scenes. The low sensation seekers had comparatively high activation in these brain regions. But when watching frightening scenes, the higher sensation seekers displayed greater brain activation in visual areas and the anterior insula than the low sensation seekers did.[21] In other words, low sensation seekers' brains weren't electrified by the terrifying scenes, rather they were relatively dampened down, while the brains of high sensation seekers got an extra charge from them.

Sensation seeking is a biological personality trait, but it changes as you age. For everyone, sensation seeking increases during childhood, peaking in adolescence or young adulthood, and then remains relatively steady until you're in your forties, when it starts to decline. These changes correlate with the horror movie fan demographic. Teens fill the seats at horror flicks, and you are unlikely to meet your grandmother there. My eighteen- to twenty-something-year-old students—from the meekest to the most self-assured—all insist that they love horror films, whereas my peers tell me they rarely if ever watch them.

Horror movies give us curios like monstrous psychopaths, vampires, zombies, and werewolves. They depict chaos, destruction, unusual and extreme forms of violence, and parades of supernatural activities, all of which increase arousal. The perverse pleasure of horror comes from the titillation that these bizarre, unpredictable, and exciting scenes bring, which itself is pleasurable for anyone with at least a moderate degree of sensation seeking.

BLOODTHIRSTY

I am not a big fan of horror movies. Perhaps you aren't either, but I bet you still dance with the macabre. I'll confess to it. Who can admit never to have craned their neck to get a closer look at the devastation after a car wreck, or for that matter tuning into the evening news, or, as I like to say, "the fire and murder hour." The crowds that public hangings drew were very rarely filled with aggrieved victims wanting revenge; rather the hordes were lured by the queer and tantalizing attraction of watching life being turned to death. Plato astutely described the seduc-

tion of death in his story of a man who passes by an executioner with a
pile of corpses at his feet. The man desperately wants to look at the pile
of bloodied bodies, tries not to, but then cannot resist giving in to his
"evil" eyes that have to stare at the "beautiful sight."[22] Why do we have
a perverse interest and even get pleasure from horror, gore, and death?

I believe that at the bottom of it all, the lure of horror is the lure of
the mystery of death. By watching horror movies, we get to experiment
with the possibilities of death in fantasy form and in such extreme ways
that we can comfort ourselves with thoughts that it couldn't possibly be
worse than the grotesque, uncontrollable, unexpected, and brutal things
that we see on the screen. Death can also be romantically glorified and
even become attractive, as in *Twilight*-type movies. Just as we are terrified by death, we are also desperate to understand it—and by exploring
the enigma of death through film we may be able to make it a little less
"horrifying."

Just as with the Sunday public hanging, our wish to discover and
conquer death underlies why millions of British viewers ghoulishly
tuned in to watch Jade Goody die in real time on national television.
Jade Goody (June 5, 1981—March 22, 2009) became famous for her
role in the British reality TV show *Big Brother* in 2002. She continued to be a centerpiece of the public eye and in 2008, when she was
diagnosed with cervical cancer, television crews and an avid audience
followed her continuously through her illness through to her death.[23]
The enticement of death and disgust also explains the popularity of TV
shows like *Fear Factor*, where challenges to consume hideous insects or
engage in daredevil stunts celebrate how one can master the horrifying
and survive.

Horror movies bring us up-close and personal with death in ines-
capable and extreme terms. From the safety of our upholstered seat,
the absurd and excessive imagery momentary alleviates our terror of
inevitable oblivion. In horror movies with "happy endings" where the
evil monster is destroyed, we even get revenge on our fears for having
tormented us. But what would it be like to live face-to-face with death
every day?

I SEE DEAD PEOPLE

I recently had the opportunity to discover what it's like to see dead people all the time—not of the ghostly type, but real flesh-and-blood dead people. Sam Greco has been a funeral-home director for more than fifty years, and I spent an afternoon in the summer of 2010 interviewing him and his wife, complemented by gin and tonics and homemade stuffed peppers.[24] It was a lovely, lively day—in contrast to the topic.

Sam started his funeral-home business at the age of twenty-five, and after burying over six thousand local residents he is now ready to retire. His wife, Catherine, herself the daughter of an undertaker, laughed when she told me, "I've lived above a funeral home my entire life." The parents of three boys, Sam answered the question that was on my mind. "Our kids were fine growing up in a funeral home. The kids never had any fear." Catherine continued, "I always told them dead people cannot hurt you. Don't worry about the dead, worry about the living." They both beamed when Sam told me that their grandson had just joined the family business, making him the third-generation descendant to service the community in death.

What is it like to make a profession of one of the most taboo topics in our society? What is it like to live with death and to deal with it daily—and not just the death of strangers, but the death of family? Sam had already buried his four older brothers. Now that he was the last one standing, was he afraid of death? And how did disgust feature in his experiences? Sam, a natural storyteller, set about enlightening me from the minute I sat down.

"I never saw a dead body in my life until I got into the business," Sam began. "I'm not afraid of death and it doesn't disgust me. You get used to it. Not that I want to rush anything along!" He laughed. "I just don't think about it." "Mind you," he continued, his voice lowering, "I had problems with embalming when I first got into it. I used to have to put gloves on. Then you get so used to doing it you can handle them just like that." He waved his hands in the air to give me a sense of his carefree attitude toward corpses. "Sometimes we make people look better in death than they did before." He chuckled and shook his head.

His difficulty in getting used to embalming seemed obvious and normal, so I pressed him on whether anything had ever really appalled him. "It's not the dead that disgust me, it's the living," he retorted. "Some of the dirty houses we go into. It's disgusting! Dog crap on the floor and there's a dead body lying there, and the relatives are just standing around. The living are more disgusting than the dead." But the incident that bothered him most was one that happened soon after he began his profession. "It was a little baby and the mother had starved it to death." Sam's mouth turned down and his eyes narrowed as he recounted the event. "I was so disgusted with her—for what she had done! That woman starved her little baby to death! She didn't have any money, she was on welfare, and she just spent her money on cigarettes and *whatever*, and never fed her own baby. That dead baby looked just like a little pigeon! Just like a little bird. Just skin and bones. When I got over there, I said to the policeman, 'This is disgusting! Why would she do this?' It bothered me so much. I scolded her, I didn't care. I said, 'What kind of person are you to let your little baby die when there's no need for that?'" Telling me the story was clearly upsetting him now.

His wife, a registered nurse for forty-two years before retiring, cut in with a change of topic to remedy the mood. "We consider ourselves caregivers."

Sam nodded. "I'm a caregiver, a protector of the dead. I want to make sure that the person's death is dignified."

I steered him onto a discussion of the supernatural and asked whether he ever felt his funeral home was haunted. "Absolutely not. I don't believe in any of that." At first he seemed slightly insulted by my question but, ever-pleasant, he winked and continued. "Sometimes my kids played jokes on their friends . . ." That would be a story for another day. I asked him if he watched or liked horror movies. "No," was his simple answer. "Why would I? I don't want to go to bed with all those things in my head. I see enough. I don't like those movies. I never did. I have no interest in it. I like serious, sensible stuff." This remark echoes other comments I've heard from people who have been exposed to the atrocities of war—they avoid fictionalized death and horror of any kind.

The most intriguing observation Sam made to me before I left for the day was how he had always noted an unusually high rate of suicides among the sons of funeral directors. Sam surmised that the suicides were because the sons couldn't deal with the pressure of the job or having to disappoint their fathers because they didn't want to go into the family business. He explained that being a funeral director spawned a lot of divorces and was a big strain on one's personal life. "Your family life is always on hold. You have to put your profession first. If someone calls in the middle of the night to tell you their mother just died, you go there right away. You have to cancel your plans or stop what you're doing—often—and no matter what time it is."

Sam might be right; a demanding, "on-call" career with a disrupted personal life seems to put one at a higher risk for suicide. The occupation with the highest suicide rate is being a doctor, especially a white male doctor. But following that are black male guards (such as crossing guards, not prison guards) and white female artists,[25] both of which are professions with rather obscure "on-call" aspects. A review of suicide rates by profession shows that the rate for funeral directors is unremarkable, and the rate for their sons is unknown if they don't already have a profession.[26] However, just as horror movies allow us to engage our curiosity with death, I wonder if continuously being around death takes away some of its terror and mystery, so that if you are seriously depressed you may see death as a logical and unfrightening option.

Evidence to suggest that familiarity with death makes suicide more likely comes from the observation that there is a surprisingly high rate of suicide among Holocaust survivors. In a five-year study of 374 elderly Holocaust survivors and 574 elderly community members who were all attending a mental health treatment clinic, 24 percent of the Holocaust survivors attempted suicide compared to 8 percent of those with no World War II experience.[27] The inscrutability of death and the fearfulness of it that those who are unfamiliar with it have may create the self-protective consequence of making suicide less likely.

AND NOW FOR SOMETHING COMPLETELY DIFFERENT

The same year that the original *Halloween* splattered blood on screens across the country, another of the most successful films ever made was splattering chewed food.

> In walks a chubby, affable guy with dark, messy hair and disheveled clothes and, to the strains of Sam Cooke crooning "Wonderful World," he merrily makes his way down a cafeteria line of assorted lunchtime comestibles. Starting with a golf ball from a discarded soup bowl, he stacks his tray and fills his mouth with nearly everything on offer, from piles of éclairs to green Jell-o squares to meat loaf, mashed potatoes, and squishy sandwiches, and then saunters over to a table of preppy peers who are less than happy to see him. Our hero sits down and, staring at the scandalized posse, proceeds to shovel a wobbling blob of Jell-o into his mouth with his fingers. "This is absolutely gross," scoffs one of the girls. "That boy is a P-I-G pig," chimes in another. And then our food lover retorts, "See if you can guess what I am now." And with a mouth full of mashed potatoes he bashes his fists against his cheeks, exploding the contents of his mouth onto the hapless group in front of him. "I'm a zit. Get it?" He grins and a massive food fight erupts that engulfs the entire cafeteria.

The star of this scene is Bluto, played by John Belushi in *National Lampoon's Animal House* (1978). *Animal House* is credited as being the movie that launched the "gross-out" genre, despite the fact that it was preceded by *Pink Flamingos* (1972), with its notorious poop-and-scoop scene. Gross-out is a subgenre of comedy movies which deliberately invoke disgust by capitalizing on slapstick and vulgar humor, an obsession with bodily functions, random destruction of property, and pleasure in the humiliation and "humorous" injury of others.

Animal House was produced on a modest $2.7 million budget, but it has turned out to be one of the most profitable movies of all time. Since its initial release, *Animal House* has garnered an estimated return of more than $141 million in video and DVD revenue. In 2001, the Library of Congress deemed it "culturally significant" and selected it for preservation in the National Film Registry. *Animal House* is ranked number one among the "One Hundred Funniest Movies" by the Bravo TV channel, and comes in at number thirty-six on the American Film Institute's list of the hundred best comedies in American cinema history. What makes disgusting so funny and such good entertainment?

THAT'S DISGUSTING BUT VERY AMUSING

Humor and disgust may seem like unrelated emotions—humor makes you feel happy and laugh, while disgust is highly unpleasant and can make you gag—but they share the same core. They are both high intensity experiences. Furthermore, laughter often erupts when a violation occurs, when something happens that isn't "supposed to," and so does the feeling of disgust. Violations can occur in all sorts of domains, from physical indignities, such as slipping on a banana peel, to linguistic peculiarities, such as bizarre accents or malapropisms, to social faux pas, such as drinking from a sterile bedpan. These violations are funny if they are harmless to you, especially if you feel distanced from the target—such as an actor in a movie—and they are funniest of all if the target is vilified—like the preppy snobs in *Animal House*. In this latter case, the violation is both mirthful and vindicating; like the monster being destroyed in horror films, the degradation of "bad guys" in comedies is rewarding and relieving. But what's so funny about watching something disgusting?

Disgusting jokes have never been in short supply, but a scientific analysis of them has been meager. In the mid 1990s, Dolf Zillmann, a horror and humor media expert at the University of Alabama, and his graduate student Patrice Oppliger,[28] had college students watch clips from *Beavis and Butthead*.[29] The students then rated how funny the clips were and completed the Disgust Scale from chapter 2. Not surprisingly, more disgust-sensitive students found the clips a lot less amusing than their stalwart peers did. Men also found the episodes funnier than women did, but even more revealing, the more disgusting the clip was, the funnier men found it to be.[30] Like sex making horror scarier, disgust makes humor funnier—to men. An explanation for this gender difference comes from the phenomenon of "excitation transfer," in which arousal from one situation augments your arousal in a different situation, but you misattribute where the arousal is coming from.[31] Since men get more charge out of cinematic sex and disgust than women do, the added energy (from sex and disgust) transfers to the accompanying emotions of fear or humor and the movie scenes are felt as scarier or funnier accordingly. The gross-out genre is primarily aimed at the teen and young adult set and has a predominantly male audience, which is the same demographic that is most captivated by disgust and fills the majority of horror movie seats. The market researchers who have discovered that the grosser the better for young males indeed have their audience pegged, and there will never be a shortage of this demographic.

Another scientific inquiry into the appeal of disgusting humor used the legendary scene in *Pink Flamingos* (1972) where the female impersonator, Divine, eats dog excrement. The aim of the experiment was to find out how one's relation to a "disgusting" event alters its amusingness and disgustingness. Participants watched the two-minute clip, in which Divine and her two sidekicks are walking down a street when they see a dog and its owner. The dog defecates on the sidewalk, and Divine sits down beside it, scoops up the poop with her hand, and puts it in her mouth. While watching, participants were told to imagine themselves either as Divine or as an uninvolved observer who had no relation to the scene. Participants then rated how they felt along six dimensions: happy, excited, angry, scared, and, most importantly, disgusted and amused.

Everyone, it turned out, was equally disgusted by watching the clip, but those who had imagined themselves as an outsider were also quite amused by it, and much more so than those who had imagined themselves as Divine. Moreover, it seemed that feelings of disgust and amusement occurred simultaneously as a mixed emotional state. That is, the viewers felt both disgusted and amused separately but at the same time, rather than the feeling merging into a blended state. A well-known mixed emotion is "bittersweet." For example, at graduation you feel happy over your accomplishment but at the same time sad that you are saying goodbye to your former way of life. By contrast, contempt is a blended emotion, where you feel disgust and superiority together as one unified feeling. It seems that disgust and amusement can become mixed emotions, and disgust gets funnier as a function of your relationship to the situation. If it's happening to someone else, it's disgusting and it's also amusing. If the disgusting situation is happening to you, it's disgusting and also often humiliating. As Mel Brooks quipped: "Tragedy is when I cut my finger. Comedy is when you fall into an open sewer and die."[32]

The scatological motif has been a consistent theme in humor since the dawn of humankind, and it is thriving today. In 2007, the animated series *South Park* was acclaimed as the funniest television show on air since its debut ten years earlier by *Rolling Stone* magazine. It has also been nominated for an Emmy Award nine times. South Park revels in gross humor, especially of the potty variety. Why is poop so entertaining? Freud believed that poop becomes funny through the process of "repression." Freud contended that because the topic of defecation has to be so greatly repressed in polite society, when it is exposed, a great deal of pent-up energy is released, like the energy inside a bomb when it is detonated. The release of this energy-tension is a relief which feels good, and the pent-up energy is converted to laughing. In other words, the more high-tension and taboo-breaking a poopy scene in *South Park* is, the more potential it has to be hilarious.

After excrement, food violations and food fights are the most common platforms for disgusting comedy. Besides *Animal House*, *Spaceballs* (1982) and *American Pie* (1999) feature outrageous food scenes. And who can forget the obese Mr. Creosote in *Monty Python's The Meaning of Life* (1983), who explodes into a million pieces after a huge repast that he finishes off with a fatal after-dinner mint?

Horror is a vehicle through which we can explore and release tension and distress about death and what disgusts us. However, horror can also be humorous. An insight I gleaned from my horror-loving students is that a central appeal of "extreme horror" is amusement. In fact, for some it is a true comedy of horrors. These horror movies are funny because of their sheer absurdity. Extreme horror is also so over-the-top that it actually allows one to become distanced and removed from the violence and slaughter it depicts. Indeed, rather than exploring death and being relieved of our anxieties of it, by watching horror viewers may be mocking death and removing themselves further from it. Paradoxically horror is transformed from a medium that disgusts us with butchery and brings us face-to-face with our utmost fear in a way that actually bolsters our feelings of immortality.

Humor is a vehicle through which we can animate and make enjoyable what makes us uncomfortable and especially that which reminds us of our animal nature, such as toilet activities, eating, and dying. Thus humor is also fundamentally about psychologically escaping or denying our mortal coils. Food is not a particularly common tool in the horror genre, but it is a winning and versatile device for disgusting humor because food and eating can be permutated endlessly into diverse and entertaining forms that elicit disgust. Food further illustrates how that which can give so much pleasure can also be revolting, and conversely how we can get pleasure from being revolted. Food is also exceptionally engaging because it is the source of one of our greatest passions. And so is sex.

Chapter 7

LUST AND DISGUST

When I lived in Toronto in the early 1990s, my boyfriend and I frequently went on Sunday afternoon strolls in search of new haunts for refreshments. One such Sunday we found ourselves in an up-and-coming part of town beckoned by a sign that read "Rodney's Oyster House." In we went. We made our way to the dark and inviting bar and ordered a round of martinis and a dozen malpeques. A large plate of raw juicy ocean and our drinks were soon in front of us, and as we picked up our glasses to clink—we paused and exchanged smirking glances—the words on the cocktail napkins revealed themselves: "Shuck me, Suck me, Eat me raw."

"Mmm," I mused. "This is going to be a good night."

Foods extolled for their ability to sexually arouse—aphrodisiacs—are intriguing because many of them are daunting to consume, especially for the uninitiated, and the challenge in swallowing them no doubt helped inspire their reputation for sexual potency. *Balut*, a partially formed duck embryo eaten on the half shell in the Philippines, hákarl, the rotted shark delicacy of Iceland, and casu marzu, Sardinia's famed maggot cheese, are all touted to increase virility and sexual prowess. Oysters can also easily be construed as hideous and onerous to swallow. Is this why they have been awarded top aphrodisiac status for centuries?

Oysters were first described as aphrodisiacs in Rome during the second century AD, and eating them with wine was purported to make women wanton. Ever since, oysters have been proclaimed to increase the libido and passion of men and women alike. Some oysters change their sex back and forth from male to female, which is where the belief that eating oysters enables one to experience both the masculine and feminine sides of love comes from. Another reason why oysters are considered aphrodisiacs is because they are sexual in shape and texture; the connection between oysters and female genitalia doesn't require much imagination. For the same reason, bananas and avocados resembling the penis and testicles respectively, are also among the list of gastronomic turn-ons.

Is there any truth to the sexual power of these foods? Bananas are rich in potassium and B vitamins, both of which are necessary for energy and sex hormone production, and some studies suggest that the banana enzyme bromelain enhances male performance.[1] Avocados and oysters are high in essential nutrients and protein, and in this way also help provide the energy necessary to perform. Regardless of any possible nutritional assistance, however, the most powerful effects of aphrodisiacs are psychological. When you're "in the mood," the body is more eager to follow. But when food is involved—like oysters—first you have to go from thinking about them as a lump of chewy mucus to sex-on-a-shell. How do we do it? And why are only some of us so persuaded? The answer is the same as with all the things where lust and disgust are intertwined—it has to do with how you sensorily experience the oysters, the situation you're in, whom you're with, your culture, your personality, your disgust sensitivity, and, most importantly, what oysters mean to you at that moment. As the Roman statesman Cicero said, "The greatest pleasures are only narrowly separated from disgust." And what separates them is our mind.

MIND GAMES

Have you ever had the object of your infatuation turn on a dime to become the object of your repulsion? I was once infatuated from afar with a colleague and for no clear reason woke up one morning feeling repulsed by both him and the fact that I'd been having sexual fantasies about him. Our obsessions can be entirely mutated by thought, especially if you haven't had any carnal knowledge of the person in question. Why? We see, hear, or learn something new and unappealing about the target of our obsession, or we have our lustful interests turn toward someone who is repaying our attention, or fickle love takes another random turn. The bottom line is that lust is in the mind of the luster. This is why so many sex self-help books extol the brain as our most important and powerful sex organ. If you aren't mentally turned on, it doesn't matter how well the hardware works.

Sometimes we are perversely attracted to people whom we actually loathe or find physically revolting. This is because lust is fueled by an animalistic passion that shares many commonalities with repulsion. Lust can also be provoked by the thought of what is obscene or forbidden. No woman would ever condone rape, yet many women have rape-like fantasies at some time in their life. Current research indicates that between 31 and 57 percent of women have erotic fantasies in which they are forced to have sex against their will, and that for 9–17 percent of these women, rape is a frequent or favorite fantasy.[2] This in no way means these women want to be sexually assaulted. Rather, rape is so far beyond what is acceptable that it becomes erotic. It is its sheer profanity that makes it sexual. What is most taboo can be most alluring, because its edgy badness gives us license to acknowledge and release the animal within us, and the animal is unfettered by societal codes. The seduction of breaking taboos is also part of what makes "affair sex" especially exciting—because it is socially condemned. We are attracted to the opposite of what we should want, often just because we *should not* want it.

Feelings of lust can be so powerful that they override disgust. In a recent experiment, heterosexual male college students were put into an erotic or non-erotic mood by first either looking at tasteful pornographic photos or nonsexual images that ranged from pictures of fully clothed women to a pointed gun. The men were then exposed to disgusting sexual or nonsexual sights, sounds, and sensations, such as a picture of a scarred naked woman or a picture of pollution, the sound of someone giving fellatio or the sound of someone vomiting, and the feel of lubricated condoms or the feel of cold and chunky pea soup.[3] Everyone found the vomiting, pea soup, and pollution experiences equally disgusting, but the men who were in an erotic mood rated the sexual version of disgusting sights, sounds, and touches as much less revolting than men who were not aroused. When you're feeling sexy, "disgusting" cues of sex are less disgusting.

It turns out that the reason lust and disgust are so intertwined is because they are neurologically in bed together, so to speak. In a recent experiment, heterosexual men and women watched erotic film clips while their brains were viewed with fMRI.[4] Among the brain regions that were especially activated was the insula, which as you've read is the same area that is most active when we experience disgust. In another study, participants were selected on the basis of whether they were sexually inclined toward S & M or not.[5] The S & M and non-S & M groups were then shown photographs of disgusting (e.g., maggots), neutral (e.g., geometric patterns), S & M sexual (e.g., bondage), or basic heterosexual sex (e.g., naked men and women) scenes while their brains were scanned using fMRI. Everyone also rated how disgusting and erotic the pictures were to them. The pure disgust and neutral pictures, such as maggots and geometric patterns, were rated the same way (disgusting or neutral) by all the participants. Not surprisingly, the S & M group rated the S & M photos as most sexually arousing, and much more arousing than the non S & M group, who found them disgusting. But, remarkably, many of the brain areas that were activated while the participants looked at photos that they considered disgusting or erotic, whatever they may have been, were the same.[6] In other words, no matter your sexual predilections and regardless of whether the photo depicted a man being whipped, a tub of vomit, or naked bodies sensually entwined, the brain was turned on in a very similar way. Not only are disgust and lust in the mind of the beholder, but for certain parts of our brain there's no difference, regardless of what our loins are feeling.

In spite of the neural dance between lust and disgust, they are not typically compatible emotions. For example, after seeing disgusting images women's self-stated sexual desire and vaginal lubrication measurably decreases.[7] Moreover, depending on what is happening in the cognitive—thinking—part of your brain, disgust can be the antithesis of lust. Imagine that while waiting at the bank for the next teller, the person in front of you turns around and licks your cheek. Now imagine that your lover caresses your cheek with her tongue. Is the first scenario repulsive and the second situation erotic? Why? They're exactly the same, except that the person doing the cheek-licking is a stranger in one case and your most intimate in the other. Moreover, you are at a bank and not in a bedroom. Just like being okay with swapping your toothbrush with your spouse but being repulsed at the thought of sharing it with the mailman, or finding a certain scent delicious in a restaurant but disgusting in an alley, it is the meaning of the situation that makes it disgusting or desirable. Factored into our own personal experience, the culture we live in is a major player in determining the meaning of what is libidinous and what is licentious.

ONE MAN'S PORN IS ANOTHER MAN'S POISON

Kevin sat alone in his office on his lunch break, his eyes fixed on his computer screen. The lusciously full breasts and pouty lips of the two naked girls rolling around on the white, bearskin rug fondling each other were getting him more excited than he had planned.

Half a day later on his lunch break on the other side of the globe, Satoshi's computer screen was playing a video where one man was urinating and another defecating into a woman's mouth while she was being sodomized, and Satoshi was getting more aroused than he'd planned.

Culture can turn the most nauseating scenarios into erotica, and Japan's $5.5 billion adult entertainment industry outranks all others for reveling in the most messy display of bodily fluids and profane acts.[8] Japan is an outwardly reserved, codified, respectful, and modest society, but beneath that surface an underbelly of unbridled sexual extremes is teeming. From erotic art where octopuses perform cunnilingus on writhing naked women[9] to films which have anecdotally sent many to the bathroom to vomit and others rushing away in tears—one blog reported this regarding a four-minute video which "involved every body fluid and looked like someone had gone to Taco Bell for lunch"—[10] Japanese pornography seems to have no limits on its imagination for the obscenely grotesque.

In a spoof exposé of the Japanese porn industry, *The Onion* claimed that leaders of the world's twenty largest economic powers signed an accord threatening sanctions against Japan if international distribution of its revolting pornographic exports didn't stop immediately. The supposed outcome was the recall of millions of depraved videos and the development of regulations that "conform to some small standard of basic human decency," which included a zero-tolerance policy for all "prurient uses" of colostomy bags and a requirement that portrayals of group sex involving seven or more must feature at least four human participants.[11]

The Japanese incident may be a joke, but in the United States various forms of pornography have also been seized and suppressed, and here it comes with questions concerning the First Amendment. You may not want yourself or anyone else to see excrement-eating à la sodomy, but "freedom of speech" implies that it is the right of individuals to choose what they want to watch, whether it makes them vomit or not, so long as it does not involve actual harm to another person. Snuff films, a shadowy evil in the porn world where women are actually murdered on camera, is a clear example of something which should be eradicated and those involved prosecuted to the full extent of the law. However, the question of whether the portrayal of women in degrading sexual situations is sufficient for censorship because of the possible harm it could do remains a murky issue.

There is reliable data that watching pornography is linked to greater acceptance of violence toward women and beliefs that women enjoy date rape. For example, in an experiment by Michael Milburn and his colleagues at the University of Massachusetts, men who watched a scene that was mildly sexually degrading to women from the movie *9 1/2 Weeks* (1986) were more likely to interpret a subsequent story describing date rape as the victim "getting what she wanted" compared to men who had first watched a non-sexual video.[12]

Degrading depictions of women vary widely, from portraying them as sex objects to eroticizing extreme violence. In fact extreme violence coupled with sex is becoming a popular new crossover genre of horror movies. Horror movies which feature sexual sadism, such as Rob Zombie's *The Devil's Rejects* (2005), are often referred to as "gore-nography" or "torture porn." Horror movies without sex feature extremely objectionable imagery, so why should sex, however peculiar, be limited by regulation but violence not? It is a notable bias that any nudity on screen requires a stricter movie rating code (R) than simple mayhem and mutilation (PG), which again shows how our society seems to object more to sexuality than to murder. Nevertheless, deciding where the line at which gore, sex, or whatever is deemed profane should be drawn leads to a slippery slope of infringements on First Amendment rights. Interestingly, in spite of or perhaps because of long-standing feminist activism against pornography, women are now becoming the fastest-growing demographic of porn consumers. Women are also increasingly filling the ranks of porn executives, directors, and producers, and taking back the night with films that cater to women's fantasies, showcasing strong female lead roles and depicting women in the throes of real orgasms.

Unpleasant fantasies such as being raped by a pockmarked colleague or a spontaneous passionate encounter with a sweaty, fat stranger on the subway dominate women's minds more so than men's. This is because women's sexuality, even in the liberated West, is more societally controlled and proscribed than male sexuality. But the erotic tension still exists and it escapes through twisted daydreams. Indeed the more forbidden sexuality is, or the more negatively certain types of sex are viewed by society, the more it backlashes in perversity. We see this when the religious or political hardliner publically beats a loud drum against some sexual act and is then caught with his hand in the proverbial cookie jar. The prevalence of sexual repression is what gave Freud his career. The puritanical culture of the Victorian period in Europe produced a storm of psychological ailments and fantasized perversity. For example, one of Freud's famous patients, the Rat Man, earned his nickname because of his obsessive fantasy that a pot of rats would be fastened to the buttocks of his father and his girlfriend and gnaw into their anuses.

The Japanese, who control their sexuality so tightly on the surface, may therefore be compelled to release this tension with the most mindbogglingly sickening sexuality imaginable. Like the explosions of laughter that excrement and other taboo jokes provoke, when tension builds there needs to be an energetic release. If we knew what porn Freud's patients liked best, we might discover that they had a lot in common with the Japanese.

Another reason why viewers keep coming back to this repulsive footage is the perverse and uniquely human tendency that Paul Rozin has dubbed "benign masochism."[13] Benign masochism means seeking out and getting pleasure from things which should be unpleasant and have no biological purpose, like watching horror movies, eating burning chili peppers, and riding roller coasters, and may also explain what propels people to want to watch or fantasize about disgusting sex. These sex acts are benign because they are only in our minds or on the screen, so we can't be harmed by them. Disgustingness without dangerousness can be sexy. We are lured by disgust even when we are repelled by it, and when we know that we aren't truly going to be mutilated or physically polluted we can indulge our Hieronymus Bosch-loving guilty pleasures.

We also know from horror and gross-out moviegoing audiences that, at least for some men, the excitation produced by seeing sex and disgust can synergize and make the experience even more arousing.

No matter what's on display, porn is big business. Five of the top hundred trafficked websites worldwide are porn sites, and sex on screen generates close to $15 billion per year in the US. This is more than major league baseball, football, and basketball combined. For comparison, Hollywood movies earn about $10 billion per year in revenue. Pornography has existed since the dawn of recorded history, when cave-dwelling artists freehanded sexually explicit romps. Along with it, there have always been people opposed to and disgusted by overt depictions of sexual behavior.

ANIMALS BY NATURE

Beyond personal ideology, culture, and erotic predilections, an element lurks beneath the sheets of pornography that makes many people uneasy. Pornography puts our animality on full, glorious display with flagrant depictions of rutting, and our animality terrifies us because it reminds us that we are squishable creatures with a finite time on this earth. Ernest Becker, the father of terror management theory, put it best: "Sex and death are twins . . . animals who procreate die."[14] Even though doing it like "the birds and the bees" sounds cute, when people are reminded that humans and animals have a lot in common they become more disgusted with the physical aspects of sex. In a study on a large class of university undergraduates, it was shown that after being reminded of their inevitable demise and then reading an essay highlighting the similarities between humans and animals, participants rated the physical aspects of sex—"tasting bodily fluids," "feeling my partner's sweat on my body"—as much less appealing than students who underwent the same procedures but instead read an essay that glorified human uniqueness.[15]

Despite euphemisms which seem to celebrate the animal–human

connection in sexuality—stud, stallion, sex kitten, foxy, vixen—the human desperation to take the animal out of sex is witnessed in all cultures. No matter how deviant the pornography is, there are always rules and limits on when, where, how, and with whom we should or should not have sex, and having sex in public willy-nilly is one pan-human proscription. Animals copulate wherever and whenever they like, but civilized humans must never (with the exception of certain isolated events like Woodstock and Tantric Hindu and Buddhist religious rituals). Grabbing your lover at the supermarket for a full-on public display of fornication will get you arrested. This literally wild behavior is abhorrent because it exemplifies the uncontrolled impulses of animals. But fear not, there are very few among us who actually want to roll around naked between the ketchup and mayonnaise jars with impatient moms and their whining children stepping by.

Just as we are most comfortable eliminating our bodily wastes privately, we are most comfortable with sex behind closed doors. We do not want others to see our animal nature. Reminders of our creatureliness stir a storm of disquiet and bring fears of our fleeting and fragile mortality to the surface. Thinking about sex in raw, animal, physical terms makes us think more about death. Despite the "sex after a funeral" motif in movies and books, which may be spurred by the desire to affirm life, thinking about our *own* death makes sex less enticing.[16] The terror management theory researcher Jamie Goldenberg and his colleagues showed that when people, especially those who are neurotic—anxious individuals who are prone to worrying about sex and death: think Woody Allen in most of his movies—were made to imagine their own death, they rated the physical aspects of sex as less appealing.[17]

In the rare instances when public sex is condoned, it is either done orgy-style or for religious purposes. In the orgy, "the crowd" becomes the social unit and the crowd is both inherently accepting of the behavior and removes personal accountability from it. It wasn't you having sex in broad daylight on the field at Woodstock; it was the *group of you*. When sex is performed before others for religious purposes, it is done with scripted rules, and sex literally becomes divine. Even so, the most

holy practitioners of these Tantric Hindu and Buddhist religions practice celibacy, because celibacy prevents animalistic urges from hijacking
their chaste souls. In nonreligious terms, we use the concept of "romantic love" to exalt sex beyond the carnal realm. Indeed, intense romantic
love feels like an exquisite spiritual union. Animals can't "make love";
only humans can.

It is ironic that though we are in denial of our animal nature, the
only type of sex that is uniformly approved of is the most basic type of
animal sex—sex for breeding. Sexual behaviors that are engaged in for
pure desire without the end result of extending the family tree have
historically been frowned upon by society, whether they be masturbation or homosexuality. Medical practitioners under Queen Victoria even
warned that sex for pleasure would lead to blindness or insanity.[18]

Nonetheless, we socially promote the alteration of our bodies to
attract sexual partners at nearly all costs, from the health hazards of
severe dieting to life-threatening mutilations. Chinese women used to
cripple themselves by deforming their feet to one-third their normal
size for the sake of "beauty." For the same reason, women in the tribes
of upper Burma ring their necks in metal and in so doing stretch them
to the point where if they were to remove the necklaces they would die
because their neck wouldn't be able to support the weight of their head.[19]
We all know that any surgery carries the risk of infection or death,
yet there is a huge industry promoting elective operations. In North
America, men undergo buttock augmentations and penile implants and
women request a variety of cosmetic surgeries from face lifts to breast
enlargements and now the latest rage; labial reduction surgery.

Labioplasty, sometimes called genitoplasty, usually involves shortening or changing the vagina's outer lips (labia), but can also include
cutting the hood of skin covering the clitoris or shortening the vagina
itself. The procedure costs at least $5,000, recovery takes an average of
six weeks, and complications include loss of sensation, difficulty urinating, and scarring. The motivation behind "designer vaginas" is the
portrayal of women in current pornography.[20] Besides being hairless and
sporting flawless bodies, the models now gracing the pages of *Playboy*

and porn videos have vaginas with very small and flat labia that reveal their clitoris. One explanation for why women in mainstream porn are shaved and shucked is so that they look less natural and therefore less animal-like (male models are usually shaved as well). The idealization of the naked human form allows viewers to forget that we are animals by nature. Paradoxically, this may also explain the motivation to nickname our lovers with animal terms of endearment. The use of euphemisms like "vixen" and "stallion" enables us to dissociate their (and our) flesh-and-blood body from the act of sex and thereby it psychologically distances us from the messy, sweaty, grunting, all-too-human-animal sex we are really having.

LOVE AND DEATH REDUX

When Lady Gaga accepted the MTV Video Music award for her hit song "Bad Romance" in September 2010, she wore a dress and shoes made of raw meat. Lady Gaga is notorious for challenging the status quo, and her award-winning song challenges not only social mores but empirical evidence on what fuels lust. The lyrics of "Bad Romance" eroticize ugliness and disease, and the video ends with Lady Gaga lying nearly naked in a charred bed beside the burnt skeleton of a customer. The visuals and words of this song are groundbreaking and attention-getting, but the message is mistaken. Death and disease don't go together with sex. In fact, disease is one of the biggest strikes against you when it comes to your success as a lover.

Being ugly is never good for attracting a mate, but just how much of a strike against you it is depends on where you live. In a global cross-cultural study with over seven thousand respondents, "good looks" in a sexual partner was rated as much more important by both men and women in countries with high rates of infectious disease, such as Nigeria and Venezuela, than it was in countries with low rates of infection, such as Norway and the Netherlands.[21] It is also the case that the higher the risk for disease where you live, the more women prefer men with masculine faces: prominent jaws, cheekbones, and brows.[22] More masculine features means more testosterone, and more testosterone means a robust immune system, which, as discussed in chapter 4, is critical for women when they are choosing a mate. Even in relatively healthy countries like Britain, just showing heterosexual adults pictures that evoke disease, such as a tissue stained with a yellowish-red fluid, increases how attractive they find opposite-sex symmetrical faces to be.[23] Symmetry is an inherent cue to good health because it indicates an immune system that successfully battled all the challenges of gestation. If disease is afoot or you're reminded of the menace of sickness by a snotty tissue, for example, finding a healthy mate becomes more important, and therefore symmetrical faces are extra-preferred. If you know that you have a symmetrical face and you're single, you might want to try having your wingman or woman drop a conspicuously dirty hanky near the object of your desire and then, when you present a charming smile, your healthy good looks could make you more likely to score. The bottom line is, healthiness leads to lust and sickliness leads to disgust—except, perhaps, if you're Lady Gaga.

Women are more disgust-sensitive than men in general, but in the arena of sex they outscore men by a large margin.[24] This is because for women sickness is a very serious peril for reproductive success. A woman's own health directly influences her chances of conceiving, carrying, and caring for a child, and the health of the man she reproduces with directly affects whether her child will grow up to thrive and have children of its own. However, no matter how healthy a woman and her mate are, they can still jeopardize the health of their child if they share genes that match up for an unfortunate recessive trait, and incest is the sexual deviance where this outcome is most likely.

Brothers and sisters and fathers and daughters share 50 percent of their genetic makeup, and not all genes work well. If a brother and sister each carry the gene for a detrimental trait that they don't manifest because they only have one copy, commingling these genes considerably raises the risk of producing a child with two copies—and therefore the defect. The incest taboo likely originated from observations of the bad outcomes sex with family members could produce.

Our immune system is responsible for the diseases we can fend off and how well, and what defects we may carry as recessive traits. How susceptible we are to the flu or whether we are born with asthma or type 1 diabetes is determined by our immune system and the genes we inherit. Our immune system is coded for by a cluster of more than fifty closely related genes referred to as the major histocompatibility complex (MHC).[25] The MHC genes are more variable than any other in nature, and though no two people, except if they are identical twins, share the same string of MHC genes, two people can end up having similar MHC genetic profiles—and the more similarities there are, the worse it is for reproduction. However, nature sometimes intervenes to protect against the transmission of lethal DNA. Unrelated couples whose immune systems are genetically similar frequently have difficulties conceiving and high rates of spontaneous abortions.[26] Either way, not being able to reproduce or bearing children who die before puberty makes you a genetic and evolutionary dead end. Is there something that women can use to help prevent this biological blunder?

THE NOSE KNOWS

No matter how good-looking a man is or how much cash is in his bank account, if he smells bad or "wrong" to a woman it is a complete barrier to sexual contact. On the flip side, if a woman finds a man's scent gorgeous, he'll have to be a complete jerk if he wants to get rid of her.[27] In research in my laboratory, we found that women rated a man's scent as the number one turn-on (or turn-off), and if a man has a scrumptious scent it is more alluring than any other physical trait.[28] Men also find a good-smelling woman to be seductive, but it isn't as important to him as how good she looks. This isn't shallowness; a woman's good looks are a critical biological indicator for men since, as discussed in chapter 4, the male goal is to have as many "potential" children as possible. The characteristics generally considered attractive in women—full lips, lustrous hair, clear eyes and skin, high energy, and especially a waist-to-hip ratio of 7:10 (think Angelina Jolie in her *Lara Croft* gear)—are indicators of a woman's fertility. For example, women past menopause are most likely to put weight on their waist before any other body part, thus bringing their waist-to-hip ratio closer to 9:10 and signaling the inability to conceive. If a woman *looks* like she could become pregnant, she is more attractive. I recently saw a photograph of Christie Brinkley in a skin-tight red dress, who at age fifty-seven had the curves of a woman half her age, though I would wager she won't be having any more children. Her superficial beauty is a misleading fertility signal; nevertheless, her excellent looks no doubt reflect excellent genes.

The reason why a man's scent is so important to a fertile woman is because our body odor is the external blueprint for the genes of our immune system. Each of us, unless you have an identical twin, has a body odor that is unique to you and you alone, just like your fingerprints. Our unique odor-print is why the search dog locates you, and not the mailman, when you've escaped from jail. When a woman finds the natural scent of a man tantalizing, her biology is whispering, "This man is a good mate because his immune system genes are *different* from yours, which will make the child you produce together most healthy." The more different two immune systems are, the less likely any negative recessive traits will manifest and the more health protection will be conferred on children, because the variation in immune system genes means greater disease coverage. By contrast, when a woman takes a sniff and thinks that a man smells clean but totally "wrong," her biology is whispering, "Get out before it's too late. His immune genes are too similar." A woman's nose is her protector against a bad biological mating.

How does this sniff mechanism operate? Does the nose contain a biological sensor that signals us when someone's body odor indicates that their DNA is good or bad for us to mate with, or does nose-intuition operate at a more conscious and psychological level? The answer isn't fully known, but it seems to be more nurture than nature. Since we can't directly test this question with humans, the next best place to turn is research with animals.

In an elegant study conducted at the University of Florida, female mice from one genetic litter were fostered and raised in a genetically different litter. When they reached sexual maturity they were given the choice to mate with (1) their real genetic brothers that they *were not* raised with, or (2) genetically dissimilar males that they *were* raised with.[29] The mice chose their real genetic brothers as mates—the ones they weren't raised with but were genetic matches to them. In other words, they used the heuristic "avoid mating with familiar-smelling mice." In this non-naturalistic scenario this was a biological mistake, but in the wild, familiar body scents would be a reliable indicator of family, and unfamiliar odors would mean mate material.

In humans, it is known that unrelated children who are raised together in communal settings like Israeli kibbutzim never marry each other,[30] and the longer two members of the opposite sex lived together when they were young, the more repelled they are at the thought of having sex together.[31] The rule for incest avoidance is, "If we were raised together, you smell like family and I don't want to have sex with you."

The smell of "family" is a barrier to sex and therefore the psychological security guard against incest. Even without a nasal mechanism to prevent sex with family members, a societal ban against it makes good biological sense. Yet incest is not a universal sexual taboo. Incest is common among the Penchama of central India, where marriage between grandparents and grandchildren is permitted, and in the state of Haryana in northern India a long history of incest, especially in rural areas, prevails.[32] Historically, among the kings and queens of Hawaii, Central Africa, the Inca Empire, Macedonia, and Egypt, incest was the typical mode of marriage and signified sacred royal kinship. In ancient Greek there was no word for "incest," and sibling marriages are believed to have been widespread during the Greco-Roman period. Cleopatra, queen of Egypt, who was Greek by descent and a formidable scholar, linguist, and politician, married her baby brother Ptolemy; her parents were also almost certainly brother and sister.[33] Incest doesn't always produce a bad outcome, unless you think Cleopatra's nose was ugly.

BAD BLOOD

Although sex is the most important activity for our biological success, the fluids that signal our reproductive capacity often trigger disgust. Semen is regarded as disgusting except under certain very limited conditions, and menstrual blood, which is the mark of a woman's fertility, is construed as the most revolting and contaminating blood of all. In the Old Testament it is written: "When a woman has a discharge of blood which is her regular discharge from her body, she shall be in her impurity for seven days, and whoever touches her shall be unclean until the evening. And everything upon which she lies during her impurity shall be unclean; everything also upon which she sits shall be unclean. And if any man lies with her, and her impurity is on him, he shall be unclean seven days, and every bed on which he lies shall be unclean."[34] Orthodox Jews uphold these customs today. In addition to the edict against intercourse during menstruation, Orthodox men must not sit on a chair that a menstruating woman has sat on, though other women may sit on it. Likewise, in Hindu cultures, anyone who makes contact with a menstruating women has to bathe, change clothes, and drink water that has been touched by gold, which is believed to be the purest metal.[35]

A number of other cultures confine women to menstrual huts and prohibit sex as well as the preparation of food for men during this monthly "curse." Beliefs about the bad blood of menstruation have also infiltrated medical doctrine, and well-respected doctors through the nineteenth century cautioned that a menstruating woman could spoil a ham and that men could contract venereal disease if they had sex with a woman during her period.[36] Anti-menstruation attitudes are not a relic of the past. In the early 1980s a large survey conducted by the

Tampax corporation reported that half of all Americans—both men and women—thought women should not have sex while they were menstruating.[37]

The accoutrements of menstruation are also considered disgusting, even fully packaged and unused. I have informally found that many men, even married men with children, either altogether refuse or find it very uncomfortable to buy feminine hygiene products for their wives. When probed further, they admit that their reluctance is due to disgust. Paul Rozin and his collaborators took the feminine protection challenge even further. Men and women were shown a new, packaged tampon, which was then unwrapped in front of them. They were then asked if they would be willing to put it in their mouths or touch it to their lips. Sixty-nine percent refused to put it in their mouths, 46 percent refused to touch it to their lips, and 3 percent wouldn't touch it at all.[38]

As you read in chapter 5, if a feminine hygiene package makes slight contact with the outside wrapping of a box of chocolate chip cookies, the cookies *inside* become much less desirable to eat. In addition to contaminating food items, feminine hygiene products can contaminate other people, as was cleverly shown in the following experiment conducted by Tomi-Ann Roberts at Colorado College, and her colleagues. Men and women college students were told that they were taking part in a "group productivity" study where they would be evaluating themselves and a partner on various merits, such as competence and likability.[39] A female undergraduate who was a confederate of the study was always the "partner." While the two students were alone filling out a questionnaire about themselves, the confederate fumbled into her purse, ostensibly to get lip balm, and with a blank expression either "accidentally" dropped a hair clip or a wrapped tampon onto the table. She then picked up what she'd dropped and applied her lip balm. When the experimenter returned, the two students were then given a questionnaire on which to evaluate each other, and the results were clear. Both men and women rated a partner who dropped a tampon as less likable and competent than a partner who dropped a hair clip. And the more

traditionally "masculine" or "feminine" the male and female participants respectively rated themselves as being, the more negatively they
evaluated the tampon dropper.

It is worth commenting that the women in both of these tampon
challenge experiments behaved just like the men. Most of us, regardless of our gender, have implicitly absorbed negative cultural attitudes
toward menstruation, which if left to fester can lead a woman to find her
own body disgusting and to avoid assessing her gynecological health.
When I was twelve, a childless and prudish adult friend of the family
died suddenly of ovarian cancer. Suddenly—because she had never in
her life been to a gynecologist because she "had no need," and therefore
had no idea that she was sick before it was too late. Excessive modesty,
embarrassment, and shame have been noted as a barrier to some women
in seeking health care in the industrialized West today, particularly
gynecological treatment, and the outcomes for health can be serious.[40]
It is troubling that these issues remain in our culture even among well-
educated and financially independent women.

As you might expect, a woman's general level of disgust sensitivity predicts how at ease she is with other people knowing that she's
menstruating and how sexually adventurous she is. Women with lower
disgust sensitivity are more open about acknowledging their periods,
tend to have sex during their period, and are more favorably predisposed toward libertine sexual activities than women with high disgust
sensitivity.[41]

MORBID SEX

Necrophilia—sexual fantasies about or actual intercourse with corpses—
would seem to be among the most vile and taboo acts, yet its practice
has been recorded since antiquity. In ancient Greece, it was typical for
husbands to sleep with and have sex with their newly dead wives as an
expression of their love for them in mourning. Legend has it that King
Herod, who had his wife Marianne executed, continued to have sex

with her for seven years after she was dead.[42] In necrophilia, just as with most other sexual fixations, individuals are typically chosen for their good looks. In ancient Egypt, to discourage embalmers from violating the corpses of beautiful women, lovely ladies were left to decompose for three to four days before being sent for embalming. That this behavior continued through history to be socially problematic is shown by the writings of the eighteenth-century Catholic theologian and later saint, Alfonso Liguori, who contended that carnal gratification with dead women should be censured as "pollution with a tendency to whoring."[43]

Not surprisingly, people with necrophiliac tendencies are drawn to occupations which give them access to corpses, namely, cemetery employee, hospital orderly, or morgue attendant. Mr. Greco, the funeral director I interviewed, told me that he had to be careful when hiring someone to make sure that they "didn't like the dead too much." Interestingly, even though many of the bodies in morgues have died from infectious illnesses, becoming contaminated does not seem to be a deterrent for those inclined toward necrophilia. Several sensationalized necrophilia cases have involved corpses whose demise involved contagious diseases, such as influenza.[44]

Clinically, necrophilia has been divided into three types: 1) "lust murder" or necrophilic homicide, where the victim is deliberately murdered so that the killer can then rape the dead body; 2) "regular" necrophilia, in which dead bodies are opportunistically used for sexual gratification; and 3) fantasy necrophilia, where corpses are not actually abused but masturbating to the fantasy of sex with dead bodies is a main preoccupation. Necrophiliacs can be either male or female, although the majority are male, and their ranks comprise homosexuals, heterosexuals, and bisexuals. Necrophiliacs are typically of normal intelligence and neither psychotic nor sadistic. Rather, the most consistent personality characteristic is very low self-esteem. The Freudian explanation of necrophilia is that it develops from poor self-esteem likely due to the loss of a significant other, possibly with additional issues such as a fear of sexual rejection, or even as a "reaction formation" to the person's fear of their own death.[45] Reaction formation is a Freudian defense mechanism where the person adopts a behavior which is the polar opposite of the anxiety they are facing. If you are terrified of your own death, a reaction formation to that fear would be to immerse yourself in death, and necrophilia certainly fits that bill.

Apart from some psychoanalytic attention, necrophilia has been sparsely studied scientifically or clinically. However, the vast amount of necrophilia in fantasy, literature, and popular culture has led some in the psychiatric community to suspect that necrophilia is much more common than presumed.[46]

Vampire myths cross the death–romance line, and many of them can be viewed as thinly veiled necrophilia. For example, the tale of "Sleeping Beauty" is considered to be a necrophiliac fantasy—a corpse is restored to life by being kissed on her lips. One of the first female psychoanalysts, Princess Marie Bonaparte, the great-grandniece of Napoleon, posthumously diagnosed Edgar Allan Poe as being a necrophiliac based upon the intense preoccupation he had with death and love in his writings, especially the poem "Annabel Lee."

> For the moon never beams without bringing me dreams
> Of the beautiful Annabel Lee;
> And the stars never rise but I feel the bright eyes
> Of the beautiful Annabel Lee;
> And so, all the night-tide, I lie down by the side
> Of my darling—my darling—my life and my bride,
> In the sepulcher there by the sea,
> In her tomb by the side of the sea.[47]

Poe's mother died when he was an infant, and Bonaparte speculated that his mother's death caused a constant unresolved mourning for her which became twisted into incestuous–necrophiliac impulses.[48] Poe's biographers, however, typically contend that "Annabel Lee" was written for Poe's wife, Virginia, who died two years before the poem was published. We will never know whether the psychoanalytical or

biographical explanation is the true one, but the poem alludes to necrophilia either way.

Freud alleged that the main function of disgust—and therefore
our disgust toward necrophilia—was to curb the "polymorphous sexuality of childhood."[49] He believed that as children we can sexually
desire anything and anyone, from inanimate objects to our parents and
animals—dead or alive. The purpose of developing the emotion of disgust, according to Freud, was so that these impulses would be curbed
and we would be equipped with a response that would help us to avoid
being sexually attracted to anything other than consenting, un-related,
living adults of the opposite sex.

In contemporary culture, many popular movies, books, and song
lyrics have obvious necrophiliac themes and a large and appreciative
audience. Besides horror movies like *Freddy vs. Jason* (2003), where
Freddy has sex with the body of a dead girl, comedies such as *Weekend
at Bernie's* (1989) and *Clerks* (2004) feature scenes where women have
sex with men they don't realize are dead. Writers including William
Faulkner have delved into the topic. In several episodes of *Family Guy*,
necrophilia is used in gags and jokes.[50] A plethora of rock bands appealing to the angst-ridden extol necrophilia in their lyrics, especially the
thrash metal band Slayer. Necrophilia captivates the popular imagination because, like watching horror movies, the idea allows us to fantasize about death at the borders of the extreme and in so doing alleviates
some of our terror of it. Necrophilia is also tantalizing because it simultaneously breaches multiple cultural taboos about sex, death, and the
sanctity of the human body, which, as mentioned earlier, is the mechanism by which many things become titillating. They become exciting
precisely because enacting them or any outward expression condoning
them is forbidden.

The fact that necrophilia is both luridly compelling and repellent,
and the inherent contradiction therein, likely plays a part in explaining why there is no federal legislation explicitly banning it in the
United States. Wisconsin was among a number of states that had no

specific legislation against necrophilia until an incident in the fall of 2006 brought the issue to the fore. In September 2006, Madison police apprehended twin brothers Alexander and Nicholas Grunke and their friend Dustin Radke, who were in possession of a box of condoms while digging up the grave of a twenty-year-old woman who had recently died in a motorcycle crash. One of them had seen a picture of the dead woman in the local obituary page and, noting that she was gorgeous while alive, decided that she would still be hot (so to speak) when dead. When the trio were brought in for questioning, Alexander Grunke confessed that their intent was to have sex with the corpse. There was no law against necrophilia in Wisconsin at the time—corpses were legally recognized only as human remains—and so the threesome were merely charged with attempted theft of a corpse.

The case didn't die, however, and in 2008 it reached the Wisconsin Supreme Court, where in a five-to-two decision it was ruled that the three should be charged with attempted sexual assault, because "[w]hoever has sexual intercourse with a person without the consent of that person is guilty of a felony." In this case, the woman in question did not "consent" to sex and therefore the three were guilty.[51] The two dissenting judges, however, have continued to argue that in legal terms necrophilia is a "victimless crime" since the victim is not a living person.

Among the states that do have explicit laws against necrophilia, sentencing varies dramatically. Nevada, home of "Sin City," paradoxically has the harshest punishment, and a person who "commits a sexual penetration on the dead body of a human being" can face life imprisonment. By contrast, in Minnesota, "whoever carnally knows a dead body or an animal or a bird" faces only up to a year in prison and a $3,000 fine.

In addition to the widely varying reactions that necrophilia evokes, from worst taboo ever to vague misdemeanor, another explanation for why there is such inconsistent and meager legislation against it is that unless there is a specific case which brings the issue to light, such as in Wisconsin, the politician who out of the blue suggests that necrophilia should be banned is going to raise more than a few eyebrows. California, where it has only been illegal to have sex with a corpse since 2004, also created the law because of a highly publicized case. In 2003, a transportation worker for a morgue sexually assaulted the dead body of a four-year-old girl who had died of the flu, and was caught on surveillance tape doing so. California currently has a maximum sentence of three years in prison for this felony, but efforts to increase the severity of the penalty are under way.

Our simultaneous revulsion against and fascination with necrophilia echoes our simultaneous avoidance of and attraction to disgust. Disgusting things, whether they be slugs, saliva, dead bodies, or sex with dead bodies, are disgusting to us, but they also *interest* us. Our intrigue with what is disgusting taps into the inherent curiosity and fascination we have with ourselves—who we are, what we are of made of, and how we will die.

QUEER EYE FOR THE STRAIGHT GUY

As mentioned several times, women are generally more disgusted by sexuality than men are, but the one sexual "deviance" that women are less repulsed by than men is homosexuality.[52] There are several reasons for this. First, gay male sex offers no biological risk for women because it doesn't involve them. Second, it only weakly challenges a woman's psychological status, because even though gay men may act feminine, they are by definition not in competition for the same mates. Moreover, gay men typically do not pose a threat of being a sexual aggressor, and for this reason women may actually prefer their company over that of heterosexual men in certain circumstances. When it comes to lesbian sex, two women in erotic union does not defile female sex-role identity because sensuality, nurturance, and being physically expressive are classic female stereotypes. Gender or sex-role stereotypes—the expected, preferred, and typically displayed behaviors of masculinity and femininity in American society—were scientifically characterized by the psychologist Sandra Bem in the early 1970s. Bem's research showed that feminine stereotypes were characteristics such as "affectionate" and "cheerful," and masculine stereotypes were traits such as "acts as a leader" and "independent."[53] Numerous experiments since then have examined sex-role stereotypes and found that most of them still hold today, especially traits like "affectionate," "compassionate," and "tender," all of which correspond to female sensuality.[54]

The mainstream acceptance of female homoerotica by women is further evidenced by the trend of young women "experimenting" with lesbian sex. In a comprehensive 2011 study conducted by the US Department of Health and Human Services in which over 13,000 men and women aged fifteen to forty-four were surveyed, it was shown that 16 percent of women between the ages of twenty and twenty-four have engaged in sexual activity with another woman, and 27 percent of all self-identified heterosexual women had experienced sexual attraction toward other women.[55]

Finally, a straight woman is unlikely to be raped by another woman, and lesbian sex has no biological consequence for her. If anything, it may decrease the number of women with whom she is in competition for desirable men. By contrast, though straight men often eroticize lesbian sex, they tend to be more disgusted by gay male sex than by any other sexual taboo.[56]

In 1895, the playwright Oscar Wilde was arrested for "gross indecency," an ambiguous charge that was used when sodomy couldn't be proven.[57] The presiding judge was so disgusted by Wilde's homosexuality that he could not even speak of the feelings that were aroused by thoughts of Wilde's acts, and gave him the "severest sentence that the law allows," though the judge thought this was not nearly severe enough.[58] The sentence was two years of prison and hard labor in a particularly harsh English jail, where Wilde suffered an injury—he collapsed from hunger and exhaustion and burst his right eardrum—which contributed to his early death.[59] Just thirty-five years earlier, a conviction of sodomy for Wilde would have meant execution by hanging.[60] Today, in eight countries, including Iran and Nigeria, the sentence for someone accused of homosexuality is the death penalty. Even in the

United States up until 2003, sodomy was punishable in most states with sentences ranging from five years to life in prison.

The irony is that in a number of regions today, such as Pakistan, Afghanistan, and Melanesia, where homosexuality is viewed as among the most despicable of all behaviors, it is not unusual for adult men to have young male wards whom they sodomize until the boys become adults.[61] As in ancient Greece and Rome, this behavior is *not* considered homosexual. In fact, it is often considered a rite of passage. In Afghanistan, where punishment for homosexuality by consenting adults can carry a death sentence, there are networks of pedophilic male prostitution rings whose clientele include wealthy businessmen, senior government officials, warlords, and police chiefs. Though illegal, these prostitution networks are widespread and referred to in Persian as *bacha bazi* ("boy play"). In *bacha bazi*, boys as young as eleven are sold by their insolvent families to powerful men and then trained to dance and sing. The boys, who are dressed in women's clothes, then perform at parties for elite men and are bought by the highest bidder to be used sexually. The sex-slave owners stridently reject the label of "homosexual" and interpret the Islamic prohibition on homosexuality to mean that they cannot *love* another man but that obtaining sexual gratification from males is acceptable. Even though these boys are tormented for years by sexual abuse, many admit that when they reach eighteen years of age, and are no longer desirable themselves, they will find boys to prostitute (males are only considered desirable before they reach the legal age of maturity).[62]

One explanation for why such widespread male–male sexual behavior occurs in Afghanistan is because male and female society is highly segregated and it is strictly forbidden for single men and women to date. Another is that Afghan society promotes male–male companionship as more respectable and valuable than interactions with women at all levels. Women are generally degraded in Afghan culture, both sexually and socially, and expected to remain virgins until they are married, where sexual congress is sanctioned for childbearing purposes only. The Afghan example shows how even a society that is disgusted by homosexuality will make exceptions for male–male sex as long as it is coded as *not* being homosexual. This again illustrates that what is sexually disgusting, what sex should be for, with whom it should take place—even a labeled sexual orientation—is a social and cultural construction. So why do straight men in the supposedly enlightened West protest so much?

There are a number of reasons for homophobia in North American culture. One is the strong presence of religious and political rhetoric that denies the "naturalness" of gay sex, and denounces homosexuality as a threat to family values. There is also a contagion dimension to anti-gay prejudice. For example, the myths that homosexuals can spread sexually transmitted diseases, such as AIDS, to unsuspecting others by the vaguest physical contact and that they can "infect" their lifestyle on unwilling, innocent victims, especially children. Therefore, antipathy to homosexuality is infused with fear of both literal and ideological contamination. Another trigger for homosexual repulsion is the basic body-fluid disgust that is evoked by thoughts of feces and semen commingling due to penis–anus penetration.[63] Curiously, however, this body-fluid interaction doesn't disgust most heterosexual men when the anus belongs to a woman. In fact, male–female anal sex is a prerequisite in nearly every heterosexual porn movie made today, which brings up what I believe is the worm at the core: power and dominance. It is the meaning of whose penis and whose anus are involved that is at issue.

Heterosexual male–male aggression finds its primary cause of violence in real or perceived threats to access to female mates, infidelity, loss of resources, and questions of paternity;[64] but a gay man poses none of these dangers. In fact, homosexuality presents no biological threat to reproductive success for heterosexual men; rather, as with women's attitudes toward lesbians, a gay man is, to a straight man, one less competitor for access to attractive women. If anything, one might argue that straight women have a biological justification to be antagonistic toward homosexuality because gay men diminish the pool of potential male mates. In any event, there is no biological incentive for heterosexual men to commit violence against homosexual men. Instead, the violence is instigated by social and psychological factors—the fact that gay men threaten masculine gender roles, as they are often identified as acting feminine, and, even more treacherously, that gay men may actually feminize a heterosexual man by sexually dominating him. The one who is penetrated is weak, submissive, and "womanly." Gay men sully the macho identity of straight men by becoming feminized in the context of sex, and elicit the private fear that any man could be thus dominated, violated, and "womanized."[65] Cross-dressing may be amusing in certain circumstances, but actually "taking it up the ass" is a serious degradation.

Sexual disgust is a social construction based on culture, context, and personality at least as much as it is a biological signal to avoid sickness or a bad mating. The modes and fashions of sexual lust and disgust spring from our ideas about reining in or releasing our animal impulses, our morbid fascinations, our questions and convictions about our own gender identity, and what we believe is right and wrong. In the case of homosexual disgust, whatever the specific reasons are in a particular instance, it can produce so much repugnance that it incites the most extreme form of eradication—murder.

Chapter 8

LAW AND ORDER

In the early morning hours of October 7, 1998, Matthew Shepard, a twenty-one-year-old student at the University of Wyoming who was not openly homosexual, was lured out of a campus bar by Aaron McKinney and Russell Henderson. The three got into Henderson's pickup truck, and before they'd gone a mile the truck pulled off the road and McKinney and Henderson began taking turns pummelling Shepard with a .357 magnum. They then drove to a remote area on the outskirts of town, where the two men dragged Shepard from the truck and tied him to a fence. McKinney and Henderson robbed Shepard, tortured him, and, while he begged for his life, beat him until they thought he was dead. Still breathing but in a coma, Shepard was found eighteen hours later by two cyclists. He never regained consciousness and was pronounced dead just after midnight on October 12.[1] The perpetrators were charged with two consecutive life sentences without the possibility of parole.

During the trial, the defense tried to argue that "gay panic" had provoked the two men to lose control and beat Shepard to death. In other words, they claimed that being so overwhelmed by disgust at a homosexual advance mitigated the crime of murder. It might have worked, but no evidence could be found that Shepard had made any moves on the men.[2] However, this defense did work for a young man named Timothy Schick. In Indiana in 1991, Timothy Schick was out drinking with his friends when his car broke down. Schick hitched a ride and was picked up by Stephen Lamie. The two men drove around for a while looking for girls, and Schick asked Lamie if he knew where he could get a blow job. Lamie told him he could help with that. The two men then went to a deserted baseball field, got out of the car, and Lamie pulled down his pants, whereupon Schick stomped him to death. The defense argued that Lamie's homosexual intent was so repulsive that a murder charge was inappropriate. The court agreed and Schick was charged with voluntary manslaughter[3]—a much lesser conviction, which in the state of Indiana carries an average prison term of ten years compared to the thirty-year average for homicide.

Matthew Shepard's story is famous for its tragic brutality, but also because after a ten year battle, it led to a significant advance for gay and lesbian rights. On October 28, 2009, President Obama signed the Matthew Shepard Act into law, amending United States federal hate-crime law for the first time since 1969, to include crimes based on sexual orientation. Had Timothy Schick been tried today, it is doubtful that his lawyers would have gotten anywhere on a homophobic disgust defense.

Our legal system is based on moral codes and we appeal to moral sensibilities when we try to justify our repugnance at sexual misbehaviors. Pee-wee Herman, a.k.a. Paul Reubens, star and creator of the multi-Emmy Award–winning children's television show *Pee-wee's Playhouse*, was arrested in Sarasota, Florida, in July 1991 for allegedly masturbating in an adult movie theater. His arrest occurred just as news of Jeffrey Dahmer being caught with a human head in his freezer was breaking. Yet Reubens's story upstaged Dahmer as a top headline for days. The crime of public masturbation, which was later resolved, instantly destroyed the Pee-wee Herman brand, and Reubens's career has never fully recovered.[4] In 1998, Bill Clinton was temporarily impeached as president of the United States because he had an affair with his intern Monica Lewinski. In Europe, men in high political office are often lauded for their Casanova-style antics. In America, we are morally strangled by sex.

THE DIRTY TRUTH

As the last chapter showed, although women are more disgust-sensitive overall than men, they are generally less offended by homosexuality. However, it turns out that, regardless of your gender, the more disgust-sensitive you are, the more revolted you are by homosexuality, even if you say you're not. In a clever study conducted at Cornell University, the Implicit Association Test (the IAT is discussed in detail in chapter 5), a test that reveals hidden feelings, was used to expose students' true attitudes toward gay people.[5] In this experiment, students' squeamish-ness was first assessed with the Disgust Scale, presented in chapter 2, developed by Jonathan Haidt and his colleagues. Then all the partici-pants were given an IAT for homosexuality versus heterosexuality. The computer keys for "good" and "bad" and "heterosexual" and "homo-sexual" were manipulated in such a way that if a participant had to think about their answer (in order to give the politically correct response) their response time would be longer than when their judgment was automatic. Thus, if someone is faster to connect "good" with "heterosexual" than "good" with "homosexual," their true feeling is that heterosexuals are better than homosexuals. The results from this experiment were clear. For both men and women, the higher their disgust sensitivity score was, the faster they were to connect "gay" with "unpleasant" and "straight" with "pleasant," compared to the alternatives ("gay" with "pleasant" and "straight" with "unpleasant"). In other words, the more often and easily you get repulsed, the more negativity you inherently feel toward homo-sexuality. Why?

One possible reason is that a heightened predisposition to be dis-gusted by homosexuals is due to fear of contamination, since various contagion myths about homosexuals are insinuated into popular culture. This is similar to the contagion explanation for disgust toward immigrants—they're unfamiliar, their customs and food are bizarre, and therefore they could infect us with "unhealthy" germs and worldviews.[6] However, other research has argued that contamination isn't the root of homophobic disgust and that it is simply body disgust—the imagery of putting "revolting" things in your mouth (the penis) and the intermingling of body fluids (semen and feces), which makes homosexual sex repulsive.[7] Whether contamination or basic body-fluid disgust makes homosexuality especially revolting likely varies between people. In any event, it seems that an aspect of homosexual repulsion is based on disgust sensitivity at a physical level.

Another factor that has been found to influence how repulsed you are by homosexuals is your political and religious ideology. The more politically conservative and religious you are, the more likely it is that you are disgusted by homosexuality. In fact, your political leaning is a good predictor for your overall disgust sensitivity. If you're a registered member of the GOP, you're likely to have a higher disgust sensitivity score than your liberal cousin.[8] Why would this be?

When we are adults, environmental factors such as our upbringing, religious ideology, social influences, income, and where we live are more likely to shape our political inclinations than our inherent squeamishness is likely to be an influence on our political leanings. Moreover, it is quite possible that our political leanings can come to modify our disgust sensitivity.

Disgust and morality influence each other bi-directionally. Morality is based on a system of values, ranging from attitudes about harm and fairness to appreciation for loyalty, respect, and purity. Anthropological research has shown that cross-culturally the moral systems of those who uphold conservative ideologies are more complex and multilayered than the moral systems of liberals.[9] Liberals are primarily concerned with two aspects of morality: not hurting others and being fair. By contrast, conservatives are swayed by all levels of morality and may place higher weight on purity, loyalty, and respect than harm and fairness under some circumstances. Could being concerned with multiple levels of morality make one more disgust-sensitive?

Conservative value systems laud order and purity more than liberal perspectives do, and since disgust is inherently about mess and disorder, your sensitivity to disgust triggers may be tempered by the degree to which "filth" at an abstract level is abhorrent to you. More specifically, it has been found that the relationship between disgust sensitivity and political orientation is most pronounced for moral issues relating to messy sexuality, such as abortion and homosexuality.[10] If your moral compass is frequently triggered, because of the multiple levels of disgust to which you are sensitive (by being conservative), then in conjunction with believing that gay marriage disrespects "family values," the fact that homosexuality also involves messy disgust triggers (e.g., feces), as does abortion (e.g., blood) means that you are likely to experience more instances where physical disgust imagery is evoked than someone who is only concerned with harm and fairness. What then happens is that the greater number of pings for physical disgust that you experience intensifies your susceptibility to disgust overall. For example, the more often you are morally scandalized, the more frequently visceral disgust may be triggered, and therefore the more likely it is for various acts and facts to become objects of your disgust. In other words, it's a snowball effect. What this means is that any morally questionable offense is liable to become infused with feelings of revulsion, and when messy body fluids are involved in the offense, one's feelings of disgust are even more vividly felt. Here is a news event that combined it all.

CRIME AND PUNISHMENT

Armin Meiwes, a forty-year-old computer technician living in Rotenburg, Germany, advertised on a website called The Cannibal Café for a man to be slaughtered and eaten. Not long after his posting he was contacted by Bernd Jürgen Brandes, who wrote in an e-mail: "I hope you are serious because I really want it. My nipples look forward to your stomach." The two men met and before Meiwes killed Brandes, they made love and then together ate Brandes's severed penis sautéed in olive oil, salt, pepper, garlic, and wine. After killing Brandes, Meiwes ate another forty-four pounds of elaborately prepared body parts over a nearly two-year period before he was caught. This incident, which has been splashed through international media, websites, books, and scholarly articles since it broke in December 2002, is legally historic. First, because there was no law in Germany against cannibalism when Meiwes was arrested. Second, because Brandes, the victim, was a successful manager with Siemens in Berlin, of purportedly "sound mind and body" when his mutilation and death occurred, and was a willing, even complicit, victim. Indeed, Meiwes had integrity in his deviance and would not kill and eat people who weren't serious about his proposition. In the time before he was apprehended, over two hundred people responded to his ad, including a chef, but only Brandes was a sincere and resolute candidate—and the only victim.

On January 30, 2004, Miewes's trial concluded with the charge of manslaughter and an eight-and-a-half-year prison sentence, because, according to the judge: "There was an agreement between them. This was the killing of a person without murder." A victim sought out his killer and a killer sought out his victim. Meiwes's defense team had in fact argued that he should be charged with "killing on demand," a lesser crime with a maximum five-year sentence. Considerable media attention led to a debate over whether Meiwes should have even been convicted at all, given that Brandes had voluntarily and knowingly participated in the act. At the opposite end of the spectrum, the prosecuting attorney was so appalled by Meiwes's actions that he continued to insist that Meiwes be charged with a much harsher sentence. Finally, in May 2006, after Meiwes had already spent two years in prison, the case was reopened and Meiwes was convicted of murder and sentenced to life.[11] This case illustrates what happens when there is a disconnect between legal justice and feelings of moral disgust, and when the prosecution can't accept that gap. Disgust changes how we view the actions of others and influences how much we think they should be punished. Feeling disgusted makes us more vindictive.

A murder where the victim is stabbed to death forty times whips a jury into a much more vengeful frenzy than if the victim is poisoned, even if the poison death is slower and more painful. Superficially clean, non-gory deaths are easier to tolerate and make us feel less morally disgusted than messy, gruesome ones. This is why lethal injection, which is the method of capital punishment most likely to inflict a barbarously cruel and painful death, is a preferred form of execution. The body isn't outwardly mangled and there's no blood. By contrast, death by guillotine, which is instantaneous and virtually painless, is viewed as abhorrent. Had Brandes been someone who simply wanted to die and Meiwes someone who had always wanted to kill someone and the two met and mutually agreed to enact their respective parts in a poisoning death, most likely Meiwes would have been convicted of "killing on demand." However, the crime in question involved homosexuality, sexual mutilation, and cannibalism, each of which by themselves can evoke visceral disgust, and when they're all put together it is almost impossible not to feel repulsed.

Our reaction to Armin Meiwes exemplifies the very serious and real way that disgust can affect us. If you want to make someone more morally severe, just make them feel disgusted. The response is easy to produce and scientifically reliable. In an experiment devised by Thalia Wheatley at the National Institutes of Health and Jonathan Haidt at the University of Virginia, a group of people selected for their ability to be hypnotized took part in a hypnosis session where they were given the post-hypnotic suggestion to "feel a brief pang of disgust and a sickening feeling in their stomach" when they read an innocuous word—either "take" or "often."[12] All the participants then read vignettes about morally reprehensible people who were involved in incidents such as stealing, incest, and bribery, and asked to decide how disgusting and morally bad the actions of the protagonists were. In some vignettes the word "take" was dropped into the description, and in others the word "often" was used. For example, "The congressman is trying to cover up the fact that he *takes* bribes from the tobacco lobby." In each vignette, when the participant's post-hypnotic disgust word was present, participants rated the story as more disgusting and morally wrong—demonstrating that subliminal feelings of disgust can be induced, and that when they are, feelings of disgust exacerbate moral condemnation. A vignette was also included that didn't have any moral consequence at all—about a student council representative who scheduled discussions on topics that appealed to both professors and students—and shockingly, when their hypnotic disgust word was present, one-third of participants also condemned it as disgusting and morally bad, even though they didn't know why. These participants expressed bewilderment at their own judgments while they were making them and offered justifications such as "It just seems like he's up to something," or abandoned logic altogether and stated, "It just seems weird and disgusting," or, "I don't know why it's wrong—it just is." In another study, people who were exposed to a commercially available fart spray, or who had to sit in a workspace that had been left in a dirty mess, evaluated moral transgressions ranging from lying on a resumé to cannibalism as more morally wrong than people in an unscented and clean work area. These effects were especially pronounced for people who tend to notice and pay attention to their own bodily sensations, or "gut" reactions.[13] The implication is that people can easily be duped by their intuitions and gut feelings, and be led to condemn people even when it is unwarranted.

There is a disturbing take-home message from this research. Jury members who feel disgusted for any arbitrary reason, regardless of whether they know why they feel disgusted, may be more likely to condemn an innocent defendant or recommend harsher punishment than is deserved. Besides having an upset stomach from the chili at lunch, gut feelings of disgust could arise from superficial characteristics of the defendant himself, such as ugliness, obesity, deformity, or foreignness, or when superfluous information is provided—e.g., that the defendant is homosexual—all of which could lead to erroneous or unduly severe sentencing. Incidental feelings of disgust can even cause people to make bad economic choices. After watching a disgusting clip from the movie

Trainspotting (1996), college students were more likely to want to get rid of a useful free gift, such as a set of markers, and to accept lower selling prices for the gift than if they had watched a sad movie clip or a nature documentary.[14] In other words, the desire to get rid of something valuable can be manipulated by incidental feelings of disgust. When we feel disgusted, we reject what is around us.

In addition to making people more judgmental and severe, disgust can simply make something "wrong." In a number of Eastern cultures, it is taboo to accept food with your left hand. Why? Because the left hand is used for wiping your private parts and therefore is near feces and urine, which are disgusting and should never be near food. We in the West don't have any problems using our left hand to take food because we don't have specific proscriptions about only using our left hand for toilet duties. Moreover, we assume that hand-washing is sufficient to rid ourselves of the dirtiness that contact with bodily fluids may produce. In some cultures soap and water are not enough. For Jews and Muslims, no amount of sanitizing can make pork an acceptable food. How do you keep people away from something that smells as delicious as bacon? The answer: by construing pigs as dirty, filthy animals, and their meat as inherently polluted, no matter what purifications are performed. It isn't only food proscriptions that can be manipulated by a little disgust trickery. As mentioned in chapter 1, anti-smoking legislation, which has banned lighting up in enclosed spaces, from baseball fields to pubs, succeeded in part by inciting moral disgust.

Not long ago, smoking was a classy, fashionable, and expected behavior. Now the message is that smoking is dirty, dangerous, and contaminating to others as well as yourself and that smokers are low-class, crude, and pathetic. Although never explicitly advertised, the emergent message that smoking is disgusting and that smokers are contaminating, socially inferior, and immoral has been a much more effective anti-smoking campaign than that smoking will kill you, which was well publicized for decades before a dent in the adult smoking market was ever made. The most powerful feature of the recent "immoral" view of smoking was conveyed by "new data" that passive or sidestream smoke had deadly consequences and was especially harmful to innocent victims, like children.[15] In other words, smokers are immoral degenerates who by their irresponsible, selfish actions are baby killers.

Driving petroleum-powered vehicles is currently a target of this type of moral disgust-infused messaging and has led to the whispered (or shouted) sentiment that people who drive Hummers are ignorant miscreants bent on carelessly destroying the planet. Being obese seems soon to be in the mix; information about how obese parents are putting their children at risk for diabetes and a truncated lifespan is becoming more widespread. It is a twisted irony of disgust that eating meat, rather than eating grubs, may be considered morally repugnant before long— "raising livestock is plundering the world's resources." If reminders of disgust and filth can make us more morally punitive, do reminders of cleanliness and purity make us more morally lenient or forgiving?

WASH YOUR SINS AWAY

"Out, damned spot! Out I say!"

William Shakespeare was an experimental psychologist way ahead of his time. In his play *Macbeth*, Lady Macbeth tries to alleviate her guilty conscience about the murder of King Duncan with a bit of hand-washing. Four hundred years later, Shakespeare's insight into human nature was proven correct. In a series of experiments, it was found that when people were first asked to recall an unethical deed from their past, they were then more than twice as likely to choose an antiseptic wipe as a free gift over a new pencil, compared to people who first recalled something good from their past (who preferred the pencil). After remembering their past treachery, people were also more likely to spontaneously complete word fragments like W__H and S__P with the words "wash" and "soap" than "wish" and "step," and to rate products like Tide detergent and Lysol disinfectant as more desirable than

Snickers bars and CD cases.[16] The moral of this story is, if you want to sell more cleaning products, remind your customers of their past wrongdoings.

In another Lady Macbeth-style experiment, undergraduates watched a disgusting movie clip and then half of them were told they *had* to wash their hands while the others weren't given any hand-washing opportunity.[17] Next, everyone rated how disgusting and morally wrong a series of vignettes were, such as eating your dead dog, stealing, killing, lying, and a man getting sexual pleasure from playing with a kitten. As Shakespeare would have predicted, the participants who washed their hands rated the vignettes as less disgusting and morally bad than those who didn't have the chance to wash away the cloud of yuckiness that hung over them from the film clip. Interestingly, the transgression that was least remedied by "cleansing" was the sex play with a kitten story—which is another example of how we are more morally scandalized by inappropriate sex than we are by killing.

Not only can washing make you more morally lax and help absolve you of past sins, the ambience of cleanliness can make you more virtuous. People who worked in a room that smelled as if it had just been cleaned with Windex were more likely to offer their time and money to nonprofit organizations than people who worked in an unscented room.[18] But sinner beware: if you've done something bad and then get to "cleanse," you can become more callous and selfish. College students who confessed a bad deed from their past and then wiped their hands with an antiseptic wipe were subsequently much *less* willing to help someone than people who revealed a bad deed but didn't get to wash up. In fact, only 41 percent of people who used the antiseptic wipe were willing to comply with a trivial request for help—volunteer to take part in a study for a desperate graduate student—compared to 74 percent of people who weren't able to first wash out the "damn spot" from their past.[19]

There is a message from this research that should concern us. We all wash daily and we are not always angels. Are people who are more scrupulous about hand-washing less kind and charitable than those who aren't as hygienic? Are the ubiquitous antiseptic cleansers foisted upon us in every venue creating a more selfish and malevolent world? While this may seem far-fetched, it's indeed possible that our hygiene obsession has created a more politically conservative society. In a recent experiment, people who were standing near a hallway hand sanitizing station endorsed more politically conservative attitudes about moral, social, and fiscal issues than people who were asked about the same topics with no Purell in sight.[20] In other words, simply being primed with a visual signal for "purity" can make us more right-wing and righteous.

Cleansing is also a prerequisite in many religious rituals. The Arabic word *wudu*, which means "partial ablution," is the Muslim procedure of washing parts of the body in water to prepare for worship. Christians follow the scripture: "Arise and be baptized, and wash away your sins" (Acts 22:16). Cleanliness *is* both literally and symbolically perceived as next to godliness. Perhaps this is why after engaging in these rituals, zealots can murder and pillage with such a vengeance. Not only have they been given a clean conscience, they now feel more hard-line about their convictions. Would a bit of dirt make us kinder?

KNOW THY BODY

Are we all influenced by manipulations of disgust and cleanliness to the same degree, or does it depend on who we are or what the situation is? And how do we engage our bodies when we are moved to wash away our sins?

> Rhiana stood in the shower for almost an hour scrubbing herself from head to toe again and again. She couldn't believe how over-whelmed with repulsion she was. It was ten years since she'd last seen him, but she'd felt his taint all over her as soon as she'd heard the news that her ex-husband had been caught soliciting sex from teenage girls online.

Rhiana is overwhelmed by the impulse to physically cleanse her-self of the tarnish from someone with whom she was once physically intimate. But there is no physical contact in the present, so what is she doing? The contamination that she feels now is through historical proxy and memory. The fact that she once accepted, loved, and touched some-one whom she has now discovered has touched others in despicable ways makes her feel polluted by his obscenities. Because of her memories of prior association with him, she has connected his current disgrace with her current self. This is a very real feeling which many of us have expe-rienced. But it is even more illogical than the eerie worry that Hitler's evilness could invade us if we put on his dry-cleaned sweater. You have nothing to do with the actions of someone over whom you've had no influence for years and are in no way socially or physically connected to, and yet it still feels as though their vile essence can be transmitted. Our memory invokes the contamination due to our past physical contact, and the more emotionally and physically involved we once were, the deeper that contamination currently feels. If Rhiana had just discovered that an old coworker had been caught for the same crime, she would be out of the shower by now.

Not only do we calibrate the level of our psychic-physical cleans-ing according to the intimacy of our relationship with the evildoer in question, we seem to intuitively know how our own bodies are involved when *we* commit offenses. We are drawn to undoing our dastardly deeds with cleansing that is related to the physical mode of our miscon-duct. When people tell a malicious lie and they *speak* it, they are later more likely to want to use mouthwash than hand sanitizer, but when they tell the same lie by *typing* it in an e-mail, they prefer hand sanitizer over mouthwash.[21] In other words, we instinctively want to clean up our dirty actions by cleansing the body part that was the messenger of our immorality.

We all differ in the degree to which we register the physical sensa-tions of our body, and what it feels like when we are disgusted. Not surprisingly, the more body awareness you have, the more you will be influenced by physical manipulations of cleanliness or filth in your moral and disgust judgments.[22] People who always notice when their stomach stars to grumble before lunch will feel more absolved by an antiseptic wipe than those who wouldn't know their foot was on fire until they looked down. Body-aware people are also more affected by simple physical manipulations of positive emotions. If you're such a person and I say "Smile!" and you do start to smile, you should be feeling a little happier right now.[23]

Our emotions can be embodied, and conversely the physical world can inhabit our emotions. People who feel physically warm, for example after briefly holding a cup of hot coffee, are kinder and more generous than people who have just held a cup of iced coffee.[24] Similarly, people who are asked to write about something in their past that they regretted and then get to seal their recollection in an envelope feel considerable relief from their negative feelings, get "closure" on the issue, and even have less memory of the remorseful event afterward.[25] But beyond any other emotion, disgust has the capacity to take hold of our body.

Disgust is literally a gut emotion. We feel our stomach roil when we are overwhelmed by disgust, and we don't need to be physically touched by anything. Likewise, many of the words we use to describe disgust—whether about contagious diseases or views on politics—refer to digestion: "sickening," "nauseating," "it makes me want to vomit." The insula, where disgust, taste, and vomiting are processed in the brain, also monitors our general internal states and physical sense of self. "Do I have an itch on my shin? Am I feeling queasy right now? How does the wool of my sweater feel against my skin? Should my arm be tingling like that? I can feel my heart beating." These examples illustrate why being sensitive to disgust is positively correlated with the personality trait of neuroticism, which refers to how anxious and emotionally reactive we are. More neurotic individuals are more body-aware and more sensitive to disgust. Disgust and neuroticism also both involve noting negative sensations. If you feel queasy and you're paying attention to feeling queasy, then you will be more likely to condemn someone or something that is in the environment with you as being disgusting—whether it's the food on the table or a robbery suspect. The defendant gets a harsher sentence when we feel repulsed because we blame him for our feelings of disgust.

CAN YOU TAKE THE WRONG OUT OF DISGUST?

Phoebe is a well-respected geneticist who just started doing research on cloning. One day while working alone in the lab, she decided to take some muscle cells from her arm and clone them in a vat. She checked on them a few days later and to her surprise they had grown into a strip of muscle that looked just like steak. Phoebe wondered whether her arm meat would actually taste like a steak, so to satisfy her curiosity she took it home and grilled it for dinner.

Zack belongs to a secret necrophilia club that has figured out how to respectfully have sex with dead people. Each member has agreed to donate their body to the club when they die and consents that anyone in the club can use their corpse for sex. A few months later Mandy, one of the club members whom Zack had always found especially attractive, dies in a car accident. As soon as her body arrives at the club he has sex with her.[26]

Can you stop yourself from feeling disgusted by Phoebe and Zack? Most of us can't help but think that their behaviors are at least somewhat repulsive and must cause some "harm."[27] Now, what if I explicitly told you that Phoebe, satisfied by her secret self-experiment, went on to productively do her regular research, never had any regrets, never repeated her taste test and never developed a fondness for human flesh. Or that in Zack's case, Mandy had no family, he was careful to use protection to prevent any risk of disease, and that following her instructions he had her body cremated immediately after he had sex with her.

With this added information, there is no logic that makes these behaviors harmful. No one was hurt either physically or emotionally:

not the people involved, not family members, not friends, not the community, not productivity at work. Moreover, these acts were private and consensual. Yet they still feel wrong. As in the case of Jurgen Brandes and Armin Miewes, who mutually consented to their respective roles in a sexual mutilation and cannibalistic killing, and where no one other than a willing Brandes was harmed, there remains an overwhelming feeling to severely condemn these acts. Why?

"Moral dumbfounding"[28] is the irresistible feeling to castigate a behavior as wrong simply because it crosses a taboo and disgust threshold, despite there being no logical reason to consider it immoral. When something feels disgusting, we search for justifications of its badness. For example, homophobics and cultural conservatives often use the argument that homosexuality harms "family values" and traditional concepts of marriage. The justification of moral wrongness then validates the emotional reaction, and it becomes a self-fulfilling loop. With Phoebe, Zack, and Armin Miewes, a line of body sanctity has been crossed—bodies were violated in ways that we consider socially and spiritually unacceptable—and if you are especially perturbed by violations of sanctity you will be more repulsed and morally scandalized by Phoebe, Zack, and Miewes. Clearly, the prosecuting lawyer in Meiwes's case was so affected, but the original judge was not.

Zack and Phoebe are fictional characters, and I adapted their stories from an experiment which demonstrated how most of us can't help but react with disgust and moral prejudice to visceral types of taboo-breaking behavior. Leon Kass, the bioethicist from the University of Chicago who chaired President Bush's Council on Bioethics from 2002 to 2005, calls this "the wisdom of repugnance" and believes that when we feel deeply repulsed by things, such as human cloning, this is a signal that a meaningful limitation is being breached and our repulsion is a valid signal that such things likely should not be allowed.[29] Many people share Kass's opinion, but as the research in this chapter has shown you, following our gut in moral instances can often lead to grave errors. Kass also chastises those who have "forgotten how to shudder,"[30] which acknowledges that many normal and intelligent people are not repulsed by certain seemingly intuitive repugnances—such as is the recently publicized development from the "Body World" exhibition.

Gunther von Hagens, the German anatomist who created the controversial "Body World" exhibition, in which plastinated male and female bodies of various ages are shown in diverse poses, is now selling human bodies and body parts online. (Plastination is a process which replaces water and fat with plastic for preservation purposes. The weight of a plastinated body is the same as that of the original corpse.) You can buy a whole body from www.plastination-products.com for about $99,463, torsos start at $79,065, and heads come in at around $31,260, excluding shipping and handling. For those on a tighter budget, transparent body slices can be purchased for about $165. However, von Hagens only allows "qualified users," who can provide written proof that they intend to use the parts for research, teaching, or medical purposes, to place an order. If you aren't a "qualified user" you can buy "anatomy glass" instead, which the website describes as "high resolution acrylic glass prints of the original body slices."[31]

It is important to know that all the people whose bodies have been plastinated gave their fully informed consent to von Hagens via a unique body donation program before their death. Only adults over the age of eighteen can sign up for the program, and the privacy and anonymity of the body donors is strictly held. To date, more than 9,000 people have pledged to donate their bodies to the Institute for Plastination in Heidelberg, Germany. Not surprisingly, religious groups have denounced von Hagens and condemned the online body shop. But the people whose bodies were sold consented to have their bodies used in this manner and therefore their personal beliefs are not being violated. Where are the limits on your moral qualms? Is von Hagens doing something disgusting and immoral? What about the "unqualified" users who want to buy his products but have to settle for "anatomy glass"?

Cannibalism, necrophilia, cloning, and the commerce of plastinated corpses all involve violations and distortions to the body and as such are viscerally repugnant to most of us. These behaviors also directly remind us of that terrifying elephant in the room—death. No matter what mitigating circumstances are given for these scenarios, it is extremely difficult for us to overcome our gut repulsion toward them. But what if a taboo were broken and no bodily violation were involved? Would it still feel disgusting, or would the primary emotion be something else?

DISGUSTED OR ANGRY?

Suppose a young white man getting onto an airplane shouts "I won't sit beside any niggers." What will the reaction to his taboo-breaking words be? Surely, outrage and fury will dominate all other emotions. We may feel some revulsion toward the man, and some sadness for what the statement represents, but the preeminent emotion will be anger, and anger elicits aggression. The man who makes this statement is more likely to be punched, thrown off the plane, or verbally assaulted than to be met by scrunched-up noses, gagging, or passengers who shrink away. In other words, he is much more likely to be negatively engaged than to be avoided. The moral insult and taboo of "nigger" elicits confrontation and approach, not disgust and withdrawal.

Herein lies the problem with the "gay panic" defense. If the murderers described in this chapter were really disgusted by their victims, they would not want to get up-close and personal and covered in the bodily fluids of the object of their repulsion. They'd want to get away. Disgust motivates us to recoil and avoid the source of our disgust, but anger provokes approach, confrontation, and attack. Therefore, in cases of homosexual murder, rage must be the primary motive, not repulsion.

Several investigations into the difference between anger and disgust have shown that we feel angry, not disgusted, about immoral acts that cause harm or injustice, such as taking advantage of someone to advance yourself.[32] Over a decade ago, Paul Rozin and his colleagues showed that both American and Japanese participants will pick the word "anger" or a picture of an angry face—from the possibilities of "anger," "disgust," and "contempt"—to denote their response to seeing someone steal a purse from a blind person.[33] In other words, anger is the emotion we feel when we react to situations that involve harm or injustice.

The differences between the situations that make us feel angry or disgusted can also be gauged by common language. Robin Nabi from the University of Arizona asked undergraduates to write about a time when they felt either "angry," "disgust/disgusted," or "grossed out," and found that 100 percent of the students who were asked to write about a time they felt angry described an event that featured being interfered with or harmed in some way.[34] But 75 percent of participants who were told to write about a time they felt "disgust/disgusted" wrote about the exact same type of anger-provoking events—being treated unfairly, offended by another's actions, gossiped about, or cheated. By contrast, 92 percent of the students who were asked to write about being "grossed out" described events involving bodily fluids, death, and insects. That is, when asked to report an event which made them feel disgust, only 25 percent of people reported viscerally repulsive events. This suggests that when we say we are "disgusted," it is in fact more likely that we are angry. The words "grossed out" refer fairly exclusively to feelings of visceral repulsion, but "disgust" it turns out does not.

New research from my laboratory has also shown that "angry" accounts far better for our feelings about nonphysical moral transgressions than visceral disgust does. We found that when people consider situations like "The man who just sold you his car lied about its condition and gave you fake contact details," they felt "angry" at least as much if not more than they felt "disgusted," but they were not "grossed out"—and this was the case no matter how disgust-sensitive they were.[35] That is, when we say, "That's disgusting," about a politician who stole taxpayers' money or someone who just sold us a lemon and skipped town, we are trying to make our feelings of moral disapproval, indignation, or outrage more emphatic, and we do not to mean that we might be about to vomit. The psychologist and author Paul Bloom similarly explains our use of physical metaphors like "lusting" after a neighbor's car or being "thirsty" for knowledge—we talk this way to make our intentions sound more forceful, but they don't actually mean we are sexually aroused by a new Cadillac or achingly dry-mouthed because we haven't read a certain book.[36] However, there are some words and expressions which do refer exclusively to physical states, and "grossed out" is one of them.

"Grossed out" means that we are somewhere on the continuum of physical disgust, as when someone has just dropped a visibly snotty tissue on the floor in front of us. By contrast, "disgust" might mean that, or it might mean that we just discovered we were lied to. Both of these situations involve violations and we reject the cause or perpetrator of them, but one feeling resides in our gut and the other resides elsewhere—predominantly in our mind—and means that we reject the person out of anger, not out of repulsion.

MAKING OURSELVES DISGUSTED

From my years of working in emotion and behavioral research, I believe that just by saying or thinking the words "that's disgusting" it is possible to make ourselves feel "grossed out." When we think about the guy who sold us the lemon and we mutter, "That's disgusting," the word "disgust" itself may trigger a conditioned pang of visceral repulsion or nausea. Our conditioned reaction of physical disgust becomes the disgust we are feeling now, though if we hadn't said or thought "disgust," our reaction would be anger. This idea fits with a long-standing theory in psychology called the somatic marker hypothesis, which was developed in the 1990s by the neuroscientist Antonio Damasio. The theory, which is mainly used to explain how emotions guide the decisions we make—as when we "just feel" someone is right for the job or guilty of something—states that when a specific emotional reaction to a situation is experienced frequently, the same part of the brain that controls the reaction when the situation is actually occurring can get turned on just by our thinking about the situation. That is, for very familiar events, our thoughts can produce the feeling of an emotion as if it were really happening, and that feeling then influences our behavior. The next time you find yourself morally incensed, try to examine how you really feel before you articulate a description of your sentiment. Then see what happens when you announce how "disgusted" or how "angry" you are. Can you make your feelings change from anger to revulsion by changing your thoughts or your words?

There is another way that moral outrage can make us disgusted—by reminding us of our death. In the airplane incident, we may feel real repulsion when we hear the word "nigger" because our perception of the world as we *want* to know it is shattered. If something is an outrageous offence against our moral codes, it can rip apart the veneer of order and stability that we work so hard to maintain and in so doing ring the bell of our mortality. When norms are broken there is disorder, and disorder reminds us that we lack the ability to control our destiny. To those for whom homosexuality is an egregious moral violation, the urge to get rid of the homosexual can be motivated by the desire to regain control of their worldview and reinforce their protective barrier against chaos and death. In the same way, someone who hears "the N word" and is outraged to the core may be motivated to retaliate violently. Indeed, it is this moral fury, and the threat to our illusion of control that it provokes, that I believe drives nearly all the violence that is committed in response to ideological or religious desecrations, such as burning flags or the Koran. Our fear of death, even when we don't realize it, is responsible for some of the worst things that we do to one another.

WHAT IS MORAL DISGUST?

As you can gather from the arguments I've presented in this chapter, a debate in morality research rages about whether our feelings about a Ponzi scheme are inherently the same as our feelings about stepping on dog poop or eating rotted meat. One experiment that recently weighed in on both sides of the debate showed that some of the brain areas that were activated when men thought about eating their sister's scab, watch-ing their sister masturbate, or burglarizing their sister's home, were the same—namely, the amygdala—demonstrating that visceral, sexual, and moral disgust share neurological substrates. However, that same experiment also showed that even though incestuous transgressions and burglary were verbally rated as equally morally bad, incest and moral situations elicited activation in very disparate regions of the brain as well as in overlapping ones. Only incest activated the insula.[37] In other words, physical disgust and moral disgust are connected, but they are not two sides of the same coin.

Another recent experiment found that the facial expression we make to the supposed origin of disgust—bitter taste—is the same as the face we make when we've been cheated out of money.[38] But anger is also triggered by being cheated, and anger activates many of the same facial muscles as disgust. Moreover, as previously mentioned, consid-erable research suggests that anger is the main emotion felt in moral transgressions that don't involve visceral or body-violation disgust, and that people simply use the word "disgust" because it has become part of popular speech to do so.[39]

While working on this book, I began to wonder about the link between moral and physical disgust and whether the connection between bitter sensitivity and disgust sensitivity was more than skin deep. To assuage my curiosity, I conducted an experiment with 162 col-lege students in which I tested their physical sensitivity to bitter taste and their emotional sensitivity to various types of disgust.[40] The way I assessed sensitivity to bitter taste was by having participants place a little piece of paper dabbed with a compound known as propylthiouracil (PROP) on their tongue.[41] PROP is medically used in the treatment of thyroid disease, and the ability to taste PROP is an excellent genetic indicator of your sensitivity to bitter and taste sensations in general. If you find a little taste of PROP as being equivalently bitter as staring at the sun is bright, then you're a "super-taster." If you think PROP tastes bitter but you can live with it, you're a "taster," and if you're perplexed as to what all the fuss is about, you're a "non-taster." Without having to take a PROP test, if you won't eat endive because it tastes horribly bitter to you then you're likely a super-taster.

In Caucasian populations, about 60 percent of people are tasters and the remaining 40 percent divide fairly evenly between non-tasters and super-tasters.[42] In African and Asian populations, the distributions are different, and there are more super-tasters relative to tasters and non-tasters.[43] Being a super-taster is due to having two dominant genes (PAV/PAV) on chromosome 5. Non-tasters have two recessive genes (AVI/AVI) at this locus, and tasters express the combination PAV/AVI. Super-tasters are physically endowed with more taste buds and other taste detectors, such as how hot they find the burn of habanero chilis and the creaminess of chocolate mousse, than tasters, who in turn are more endowed than non-tasters. The bottom line is that everything in the mouth of a super-taster feels and tastes more intense than in the mouth of a non-taster.

Participants rated their response to PROP using a standard scale for this type of research, which has verbal labels ranging from "no sensation" to "strongest imaginable sensation of any kind." From their scores on this scale I categorized the participants as tasters, non-tasters, or super-tasters, and then examined whether their sensitivity to bitter taste was related to their sensitivity to sexual disgust, such as "hearing two strangers having sex," body and disease disgust, such as "seeing an unflushed bowel movement" or "sitting next to someone with red sores on their arm," and moral disgust, such as "stealing from a neighbor." What I found was that, overall, super-tasters were more disgust-sensitive than tasters and non-tasters, and tasters were more disgust-sensitive than non-tasters. But when I looked at responses to the three subtypes of disgust a specific pattern emerged.[44] How bitter PROP tasted directly correlated with how disgusting "hearing two strangers having sex" and "seeing an unflushed bowel movement" were rated to be—super-tasters were most disgusted, non-tasters were least disgusted, and tasters were in the middle. But when it came to moral issues, such as "stealing from a neighbor," taste sensitivity had nothing to do with disgust ratings. That is, your response to eating endive has no bearing on how disgusted you would feel by being sabotaged by a colleague. Moral disgust is not related to bitter taste sensitivity, but physical disgusts are. What does this mean?

First, we are born being a super-taster, taster, or non-taster. Second, both disgust and taste are processed by the anterior insula. Therefore, super-tasters have gone through a lifetime of experiencing more oral sensations that intensely activate their insula than tasters, and tasters have undergone more intense insula activation than non-tasters. My speculation is that the insula becomes sensitized as a function of one's life history of oral sensations, which causes a neurological ripple effect such that other experiences which are governed by the insula (e.g., disgust) become more sensitized. This is why super-tasters are more reactive to disgust than tasters and non-tasters. Just as having more pressures on your moral barometer may make you more disgust-sensitive at a psychological level—which explains why conservatives are more homophobic than liberals—sensitivity to bitter taste may make you more disgust-sensitive at a neurological level, and explains why super-tasters are more repelled by seeing an unflushed bowel movement than non-tasters are.

The debate over whether moral disgust is part of the same constellation of disgust reactions as tasting something bitter, seeing overflowing toilets, and thinking about necrophilia is far from over. My analysis, however, is that we only feel "grossed out" by moral transgressions when they involve the body, and though moral and physical disgust may feel similar in certain instances they are not the same in kind. Rather, where morality and disgust are connected is in the way that we respond to them.

PURE OF HEART AND MIND

Consider the scenario of stepping on dog poop while taking a stroll through your favorite park. You feel physical disgust at being immersed in animal waste and you also feel moral disgust at the irresponsible lout who left her dog's poop in the middle of a public walkway. Where these two feelings come together is in avoidance—and, more to the point, in rejection.

Just as we want to keep away from poop or a stranger's oozing sores, we reject the immoral person who is not upholding her end of the social contract. These selfish rule-breakers are bad for the social health of the community as well as for our personal welfare. The immoral person might exploit you, steal from you, and make your community unpleasant or unsafe. More abstractly, but just as ominous, being around the badness of immoral people might sully our inner, spiritual self or soul. Like the creepy feelings of contamination that arise when we are asked to put on a sweater that a mass murderer owned, the fear of contamination from someone's depraved spiritual essence arises when these people infiltrate our society. Our physical body can be contaminated by disease, and our social community, psychological self, and soul can be contaminated by moral turpitude. Disgust at maggots on meat prevents us from eating rotted food and incorporating sickness into our body. Disgust at the beliefs and behaviors of those who threaten our spiritual good health prevents their moral sickness from contaminating us psychically. The connection between physical disgust and moral disgust is that they both help us to avoid what might corrupt us, and it is in the rejection of these people and things that disgust and morality intersect. Moral revulsion motivates us to reject impure and corrupt people, and disgust provokes withdrawal and rejection of bodily threats.

The more you contemplate your inner spiritual self, the more vigilant you will be about the possibility that your soul could become fouled. Being repulsed at the possibility of contamination by evilness, foreigners, homosexuals, or scoundrels is not a fear of physical disease, but a fear of psychological or spiritual ruin. The reason why you want others to endorse your beliefs and behave as you do is because you don't want to be spiritually blemished by having to interact with people who are morally corrupt, whether they believe they are or not.

Our concern for our soul falls back on our fear of death, and this fear is lessened by believing that something about us is immortal. If the eternal part of you is your soul, then you need to ensure that its interminable future is the best it can be. In all religions, the things we do on the earthly plane determine how pure or stained our soul is, and this determines the pleasure or pain of our afterlife. It is the simple human desire for happiness that motivates us to keep our soul as clean as possible so that it can experience a perpetuity that is as pleasant as possible.

Atheists and people who don't believe in a soul or spiritual afterlife also cling to concepts of their immortality. There are those, like me, who take comfort in the belief that the atoms that compose me will be recycled through the universe forever. Others are solaced by thoughts that their children and genetic lineage will perpetuate endlessly. Another reassurance is in the belief that the artifacts and ideas we create will be eternally present in the social-intellectual world. Leonardo da Vinci has been dead for nearly five hundred years, but as long as his art survives so does his essence. The emotional reaction many people experience when hearing about vandalism against the *Mona Lisa* is as much a response to the injury it inflicts on the spirit of the painting and Leonardo as it is about the attempted desecration of an object with high cultural value.

We feel disgusted when immoral acts violate our sense of what is good and pure—and therefore negate what civilized humans are supposed to be like. By being repelled by beliefs or behaviors that threaten "humanity," moral disgust functions to keep our social communities, psychological self, and soul clean and away from contamination. Rejection of dirty food, dirty people, dirty thoughts, and dirty behaviors helps keep our bodies and our souls alive and well.

Chapter 9

DISGUST LESSONS

What can we learn from disgust? Can we harness the intricacies of disgust to make us happier and kinder? Can we use disgust to win friends and influence people? What can disgust teach us about ourselves? Can we deconstruct disgust as it is overtaking us and stop it from hijacking the moment? Might there actually be a bright side to disgust?

Most likely you have already gleaned some tricks and tactics from the previous chapters on how to use disgust for your practical and personal benefit. But in case you haven't been scheming, here are a few tips. If you want to assuage your guilt after committing a disgusting moral act, wash your hands or use some mouthwash, depending on what part of your body carried out your sins.

If you see that you are about to be attacked and you are clearly weaker than your assailant, try to release some bodily fluids, the grosser the better; tears won't work. Unless he's a psychopath, disgusting your assailant will make him recoil from you and could give you the opportunity to escape. Similarly, if you want to hide or safeguard something from being rummaged through, you can protect it in something that would be disgusting to touch, like a soiled tissue, chewed bubble gum, or a used vacuum cleaner bag. I'm sure you can think of more.

If you're a lawyer, you can use the power of disgust to manipulate the defendant and courtroom in order to alter the conviction and severity of judgment in your favor. If you're looking for leniency, present your defendant looking as clean and respectable as possible and offer the jury and judge some hand wipes or, better yet, warm, moistened washcloths, before the session begins. Your defendant's crimes will seem less heinous and the jury and judge will feel warmer and be more compassionate and less severe in sentencing. It's more difficult to be subtle about manipulating for prosecution, but you might casually tell a disgusting joke before the courtroom session starts, or surreptitiously litter the jury box with discarded food wrappers, or make the judge's table sticky. Disgust and filth make persuasively unfavorable impressions on our moral judgments, but also beware that beauty confers unfair advantages. Ted Bundy was a hard sell for the prosecution because he was so good-looking and "clean-cut."

If you're a politician with conservative views, you can use disgust to manipulate your voters. New York state Tea Party activist Carl Paladino did just that as a gubernatorial hopeful in the 2010 primary elections and trounced his conservative rival, Rick Lazio, by a whopping 24 percent.[1] How did he do it? Paladino mailed campaign flyers to 200,000 registered Republicans in New York state that were impregnated with the aroma of rotting garbage. In the malodorous flyer Paladino railed against Democrats who "betray the public trust" and pledged to "end the stink of corruption in Albany." The flyer was headlined with photographs of seven Democrats, six of whom had been investigated, including the previous New York governor, Eliot Spitzer, who was forced to resign over his involvement with prostitution. The smelly flyer alone can't account for Paladino's whopping win, but the disgust induced by the putrid pamphlet would have elevated the moral condemnation of the recipients and made them more eager to endorse someone who promised to "ferret out corruption" and "get rid of the stink."

Notably, Paladino didn't beat his Democratic opponent, the current governor of New York, Andrew Cuomo, in November 2010. Perhaps voter booths had been aromatically disinfected and the "clean"-feeling voters, like the people who worked in a room that smelled as if it had just been scrubbed with Windex,[2] were inclined to be more liberal toward social causes. Take heed, however, that by the logic of my fanciful voter manipulation it could have turned out differently if, rather than a clean-feeling scent, hand sanitizer was visible and available to voters before they went to cast their ballots. If hand sanitizers were stationed beside ballot boxes, voters might be primed to think of moral purity and traditional values, which in turn could favor wins for the Republicans.

DISGUST DOESN'T SELL

> Shari walked into a convenience store in downtown Toronto and asked the checkout clerk for her favorite brand of cigarettes, but when he handed her a pack she grimaced. It was emblazoned with a full-color photo of a bloody bisected brain and the tagline "Warning: Cigarettes Cause Strokes." Shari held the pack away from her and asked for a different one (Canadian cigarette packages are randomly assigned anti-smoking images). The clerk smirked and handed Shari a pack with a picture of a bar graph and header that read "Warning: Each year the equivalent of a small city dies from tobacco use." Shari smiled weakly and paid the $15.75 for her one package of twenty-five cigarettes.

Fifteen graphic anti-smoking warnings have marked Canadian cigarette packages since 2000. Many other countries also have very prominent and gruesome anti-smoking messaging on their cigarette packages—most notably Brazil—but up to now, the US has had among the least prominent warning labels of all. This is soon to end, as the Food and Drug Administration has recently chosen nine warning labels—which include a picture of a dead smoker and blunt warning words such as

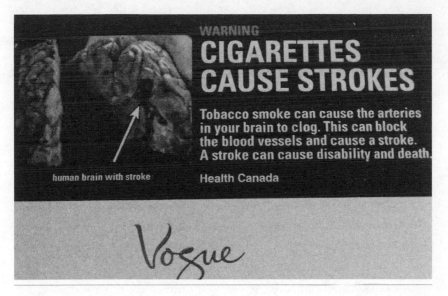

Figure 9.1

Canadian cigarette package with warning

"cigarettes can kill you"—that will cover 50 percent of all cigarette packages sold in the US as of September 2012. But the question remains:
do disgusting pictures of bloody brains, mouths festering with cancer,
or even corpses stop people from smoking? Does having to pay more
than fifteen dollars for a package of cigarettes, as is the case in Canada,
stop them? Although Canada has seen a recent decline in the number
of reported smokers, it is not known why. Smoking is on the rise worldwide, especially among young adults.[3] High cost may deter some, but it
has also increased black-market racketeering. The Canadian campaign
of disgusting imagery has not clearly tracked a drop in smoking, and
this is because the problem with gruesome imagery is that it just makes
people not look, it doesn't make them not buy what they are addicted to.

If an image is disgusting people will avoid it—they look away and
block out any information that comes with it.[4] Because Shari took the
package with the ungory graph, she may actually read the warning and
absorb the information it displays—that "each year the equivalent of
a small city dies from tobacco use."[5] This fact may have some influ-

ence on her smoking, though teens and young adults are rarely fully convinced of their mortality. As you read in the last chapter, the best way to make smoking unpopular is to make the act itself uncool and debased. Casting young smokers as "disgusting" social pariahs will do much more to deter the use of cigarettes than disgusting them or using scaremongering tactics. This is why the moral condemnation of smoking that emerged in the 1990s was so successful.[6] Social disapproval of smoking is high in most parts of Canada; some cigarette warning labels emphasize that smoking can kill innocent victims, such as children, not to mention being a cause of impotence—another major motivator. It is most likely that a combination of social, moral, and financial reasons has led to the current trend in Canada, because there is no evidence that gore is a deterrent to those who need to light up.[7]

Besides being ineffective at curbing unhealthy addictions, disgusting images do not motivate people to donate to charity, even though their use is widespread in charity advertising. There are more than 800,000 charitable organizations in the US, and they are all in competition for your hard-earned dollars. To win your generosity, nearly $8 billion per year is spent in marketing in the US alone. The obvious point about how that money should really be used is noted. From an economic perspective, however, the question is, what actually works?

TV ads that feature ragged, sick, dirty, fly-bitten children are not effective at eliciting charity because the images are too aversive. Disgusting scenes make people look away or change the channel to avoid them; they do not engage people and make them want to reach out to these deplorable victims. By contrast, showing sadness increases charity. In a recent study conducted by Deborah Small and Nicole Verrochi at the Wharton School of Business, flyers were given out to a large group of participants ranging in age from eighteen to forty-three. The flyers ostensibly advertised an organization seeking aid for research on childhood cancer, and showed a picture of a healthy child making either a sad, happy, or neutral facial expression. Seventy-seven percent of the participants who got the flyer with the sad-faced child donated, compared to approximately 50 percent of the participants who got flyers with the neutral or happy faces, and their donation amounts were almost twice as large. This study also demonstrated that, like disgust contagion, people can "catch" the emotion of sadness and are particularly sympathetic and likely to donate when they see a sad-faced victim.[8] This strategy works for animal charities as well. "Sad" expressions on dogs and cats are more effective than scenes of abuse at capturing people's generosity.[9]

In addition to displaying expressions of unhappiness, research suggests that charity portraits should show their victims looking as clean and healthy as realistically possible. If the victims are depicted in disgusting conditions, even if this represents reality, it may backfire by making the viewer disgusted and hence turn away and turn off.

You can also use the converse of disgust—emotional cleanliness—to manipulate people to be more giving. Aromatic signals of clean have been shown to encourage charitable generosity.[10] Of all our senses, scent is the most powerful for eliciting emotion, because of its uniquely intimate connection to the neurological seat of emotional processing, the limbic system. Therefore, if you are organizing an important fundraiser, you would be well advised to use "clean" scented products to prepare your meeting room, as your guests are likely to become more philanthropic. Despite intense efforts, a way to transmit scent through television has not yet been effectively developed, but charity flyers that are impregnated with clean, fresh scents might help induce magnanimous donations, just as stinky flyers can intensify judgmental and conservative responses.

You can also enhance donations to a charity by reminding people of their mortality. For example, presenting an ad for funeral arrangements or life insurance before the main attraction might be a good idea—but only so long as your charity doesn't also remind people of death. In other words, these types of manipulations would be good for soliciting money for Habitat For Humanity, but not for a cancer research society.[11] However, if your organization does inherently evoke a reminder of mortality, an effective approach would be to take advantage of occasions when potential donors are high in self-esteem and their worldview is lifted—for instance, collecting money for the Jimmy Fund after singing the national anthem at a baseball game.

GET OVER IT

> Anika took her boyfriend, Roy, for the walk in the woods she fol-
> lowed on her bi-weekly dog excursions, where busy people paid
> her to work out their pets and give them fresh-air freedom. Anika
> knew there were lots of harmless snakes in the forest; she'd been
> confronted by them many times. At first, she was so repulsed that
> she almost changed her route, but inertia got the better of her
> and by now she'd seen so many that they didn't even get a flinch
> out of her. As they rounded a turn by a small swamp, Roy suddenly
> leaped back and yelped, "Oh my God! Get me out of here! Snakes!
> There are snakes everywhere!" Anika laughed at him and chided,
> "It's nothing, they're harmless. Let's go, Big Guy."

Are there any items on your disgust list that you'd like to cross off? You know they're useless and impeding your potential, but you can't seem to get over them? If so, the best way to undo, or at least reduce, your disgust reaction to specific triggers is to overexpose yourself to them. Overexposure makes the triggers lose their power. Anika was initially repulsed by the snakes too, but now she is indifferent. I used this method to get over my disgust for the elephantiasis victim. I looked at the picture so many times that I got used to it and it didn't bother me anymore. Likewise, medical students discover that they have become much less disgusted by death and gore after they've dissected a cadaver in class than they were beforehand.[12]

The classic technique that clinicians use to help people overcome their phobias—systematic desensitization—is safe to try at home. If snakes are a phobic disgust trigger for you and you'd like to do more walking in wet woodlands, you can start by looking at line drawings of harmless snakes while thinking calm thoughts. When you can look at these illustrations without grimacing, move up to more lifelike photographs. Surf the web for the types of snakes inhibiting your strolls and then stare at them until you can look at whatever pops up on the screen without cringing. At last, when you're feeling brave enough, take that walk. It might be helpful to bring a friend who isn't disgusted by snakes and to have a relaxation mantra in your head. Deep breaths. When you do come across one of these wriggling critters, ask yourself why you're disgusted by them and try to de-disgust them.

I've just reviewed a few ways in which you can wield disgust and its offshoots to influence others and yourself, and you can probably work out additional tactics that would be personally useful to you. But what have we learned about the deeper meaning and purpose of the emotion of disgust?

THINGS I'VE LEARNED SO FAR

I hurried onto Boston's Back Bay train platform, trying to decide where to position myself for best access to an uncrowded compartment when the train arrived. As I surveyed the density of various clumps of waiting passengers, I noticed a wiry, peculiar-looking young man sitting between two elderly people. As I approached, his spastic movements and twisted arms came into view and I felt myself recoil as I deciphered his palsy. But I was also curious. What was the situation? What was the relationship between the trio? Grandparents and grandson? Ward of the state and volunteer workers? I strode by and stared directly at them and at that moment was slapped with a fistful of revulsion. The young man had his left eye sewn shut, and the patch-up job was obviously recent as big, bloody stitches crisscrossed his eye and over his pale and bruised face. I was so appalled that I am sure I lost the social ability to control my expression. Embarrassed and sickened,

I quickly moved as far from the trio as I could, only to discover when the train pulled in that I had to walk back toward them in order to board.

I am ashamed to tell this story of my antipathy, but it was a profoundly insightful lesson in disgust. First, I discovered that if I had been prepared for seeing the man and especially his eye, I would have been better able to control both my outward expression and inward emotion. This is why I was not so disgusted by being the nose judge at the Rotten Sneakers Contest. I had prepared myself by thinking over and over about having to sniff those smelly sneakers and was braced for the worst. When the time came, being the odor judge was not nearly so bad as I had imagined. From my train platform experience, I also realized that the shock value of the incident was key to its emotional potency. Surprise, as the flip side to preparation, is the great augmenter of disgust. Without surprise, many things are much less revolting. This is why horror movies always use shock gimmicks. If you know that a bloody hand is going to lurch from the ground it is a lot less horrifying, just as if I had known that I was about to see the man's bloody eye it would have been more tolerable.

I further pondered why a maimed eye—the window of our soul—is so disgust-inducing. Of all our exposed body parts, the eye is the only one that is moist and gelatinous, which is itself a trigger for disgust. Our eyes are also critical to our human, vision-centric mode of existence. Eyesight is what people who have all their senses intact think would be the worst of all senses to lose. In fact, losing the sense of smell is often far worse; because of its intense connection to our experience of emotion, the depression and loss of sense of self that can occur with smell loss can be utterly devastating.[13] But besides being our portal to vision, our eyes are also special because they cover two of the holes on our face, and we are extremely protective of our holes. Our anus, genitals, mouth, nostrils, ear canals—these are places where the outside world can breach our bodies and bring contamination and danger inside of us.

Paul Rozin has shown that we are much more perturbed by the prospect of a clean poker chip touching our mouth than if something touched our elbow. Contact with the inside of any of our undefended orifices is also worse than contact with the entryway to our insides. For example, we are more disgusted if a Q-tip touches our tongue than if it touches our lips.[14] The eyes cover the holes behind them, but they are also very delicate and can be so effortlessly maimed that we can easily be invaded through them. Moreover, the vulnerability of our eyes to mutilation reminds us of our inherent animal frailty, and brings our mortality to the forefront of our mind.

Another revelation from my train platform experience is that when I had reached a safe distance away from the man and his caretakers, I kept stealing glances back at them. What had happened to his eye? What was that couple doing with him? Were they taking care of him, or were they somehow the cause of his misfortune? I was enticed by the trio in the same way as I am lured to peer at the wreckage of a car accident, just as when I was a child and became obsessed by the photo of the man with elephantiasis. I was "ready" to be disgusted, and something happens to us when we are ready for disgust: we become intrigued by it. Disgust is fascinating because it taps into our curiosity about the ultimate mystery, death. By putting a toe into the icy waters of gore and deformity, we are sneaking a peek at death and subconsciously contemplating, how will it happen to me? This led to my key revelation: disgust is all about me.

EMPATHY'S TWISTED SISTER

Of our six basic emotions, disgust is the most egoistic. All emotions involve *our* reaction to a situation. Happiness, sadness, anger, fear, and surprise are directly to do with what we are experiencing—a win, a loss, an encroachment, a threat, a shock—but they are much more automatic, reflexive, unmeditated, and outward-focused than disgust is. Disgust requires a degree of egocentric thinking and self-awareness that young children and animals do not possess, and the unfolding nature of disgust further enables the more elaborate cognitions that are necessary for disgust to take place. Disgust is intrinsically self-focused and self-reflective. In fact, I have come to believe that disgust is about empathizing with yourself.

Empathy is a complex social emotion. We experience empathy when we feel and share the emotional experience of another person. But that "other" can also be us. We can think about an event in our future and imagine ourselves in a certain emotional state and feel it right now. If you seriously contemplate what it would feel like if in five years you discovered that your spouse was cheating on you, you can feel hurt, lost, angry, and so on—at this very moment. You feel and share the emotions with yourself in the future, even though the event isn't happening and you may rationally never expect it to. Disgust involves a similar form of mental abstraction.

We empathize with ourselves at the potential for having a contaminating or repulsive experience, or egoistically with ourselves in the moment as the object touches us. The things that incite our self-empathy are the things that disgust us. When you see the snake slithering near your foot, you feel disgust because you empathize with yourself for what the possibility of being touched by the snake would be like. When the snake is actually slithering across your bare toes, the loathsome sensation that fills you is self-empathy because of what the snake means to you right now—it's disgusting!

Our feeling of disgust when we walk past a bloody eviscerated squirrel is not because we are empathizing with being a squirrel, but because in the recesses of our mind it elicits the thought that we too could be smeared dead across the pavement. Likewise, if you are disgusted by a person who is nothing like you, you are in fact implicitly empathizing with the possibility of being like him or her: horribly obese, ugly, covered with red sores, a homeless drug addict. I saw the man with the bloody stitched-up eye and I turned away in horror because I understood that his condition could happen to me. And the more you are in fact similar to the object of your egocentric empathy, the more intense the feeling of revulsion is, even if the connection is very superficial. For example, we tend to empathize more with members of our own sex than with our opposite-sex peers.[15] If the person at the station had been a woman, I would have been even more repulsed.

Another way in which disgust and empathy are selfish is that they enable us to provide help to others because that helps ourselves. You see my expression of disgust after I smell the milk and now you don't try it, or your suffering makes me feel bad and so I am motivated to soothe you. My reaction has positive consequences for you, but I benefit first, and if I didn't benefit I would be much less likely to help you. Even people who rush into a burning house to save a stranger, or do other heroic deeds, nearly always feel very good inside themselves for what they have done, which is a reward in itself. Pure altruism is very rare.

Disgust at purely sensory experiences, like the stench of sour milk or human waste, is egoistic as well, because to be disgusted by it you have to know what it means in relation to you. My PhD adviser used to joke about how people like to smell their own farts (my research was about smell). Unless you hate yourself, your farts don't disgust you because they just mean *you* and whatever you ate. Someone else's farts are disgusting because they represent possible contamination by an outsider and their foreign waste. The only situation in which your farts might truly disgust you is if they portend sickness, in which case you are again empathizing with yourself for the distress of being ill, on top of which, disease itself is a fundamentally forceful trigger for disease.

Disgust is the twisted side of empathy and egoism. It has not been scientifically tested, but I wager that the stronger your self-preservation instinct is, the higher your disgust sensitivity will be. Disgust is a survivalist emotion for the self; therefore, the stronger your disgust motive is, the more likely you're going to survive and have children and pass on this reactivity. This is why the impulse for disgust is so strong. It is highly adaptive.

I FEEL US IN MY INSULA

In chapter 3, I discussed an experiment which showed that the same activation in the anterior insula was produced when people watched a video of people smelling a disgusting odor and when they smelled the same odor. This finding demonstrated that we feel other people's disgust as our disgust, and that disgust empathizing takes place in the same part of the brain where disgust is processed, the anterior insula. An even more impressive experiment that underscored the connection between disgust and empathy was done with taste.[16]

At the beginning of this study, participants' level of empathy was assessed by a standard questionnaire. Then, while the participants' brains were being scanned, they watched a video of people sampling bitter and sweet liquids, and then they tasted the same sweet and bitter solutions. The brain scans revealed very similar activation in the insula when the participants watched another person taste bitter and sweet and when they themselves tasted bitter and sweet. This means that it isn't just disgusting sensations that activate the anterior insula; empathic reactions to positive (sweet) experiences activate the insula too. Of course, the anterior insula is also where taste in general is processed, so the result may not seem too impressive. But strikingly, the higher the participant's empathy score was, the brighter their insula glowed when they watched other people taste both sweet and bitter solutions. Feeling disgust and feeling empathy for both good and bad experiences activates the insula, and the more empathic you are, the more the insula is turned on. Empathy and disgust overlap in the brain.[17]

The insula also feels our inner self, and self-awareness is crucial for empathy, as well as for disgust. Most empathy experiments have been done with pain; you watch someone getting painful shocks, pin pricks, hot or cold sensations, and your empathy for their pain is examined. In these experiments, after watching someone else suffer, you usually get the fun of experiencing the pain as well. This research has consistently shown that the same neurocircuitry—principally the anterior insula—is involved when you personally experience pain and when you see another person undergo the same painful procedure. The only difference is in degree.[18] There is more intense activation when the pain is happening to you. Empathy, like disgust, is a selfish emotion, and it depends on self-awareness at least as much as "other" awareness. This is why in our insula, the representation of our self and the selves of other people is shared.[19]

We also need empathy in order to make moral judgments. You have to be able to empathize about what it would be like to be harmed or have your rights taken away in order to know that these actions are morally wrong. Psychopaths have no moral conscience, and they have no empathy or emotional disgust. Compared to psychologically healthy people and even other violent offenders, psychopaths have reduced insula activity and insula size. More amazing still, another group of people who can't recognize disgust also lack empathy—people with Huntington's disease. Huntington's disease patients have damage to their insula, and research has shown this damage compromises not only their understanding of disgust, but their ability to feel empathy as well.[20]

If you're a natural-born psychopath, you started developing traits such as conduct disorder (defiance of authority), indifference to other people's suffering, and aggression during childhood and adolescence.[21] This is also the time when most people develop empathy and learn disgust. Studies on adolescents who are at risk for becoming psychopaths have shown that their anterior insulas are smaller than those of normal teenagers, and the more aggressive and the less empathetic the at-risk adolescent, the smaller their insula.[22]

The "gore watcher" personality among young horror film fans also fits with my theory that empathy is needed in order to feel disgusted. Gore watchers identify with the villain–monster and do not empathize with the victims whose blood and guts are being splattered on the screen, nor could they be truly disgusted by the gore or they wouldn't seek it out. I expect that a predisposition to psychopathy, which implies having low disgust and low empathy, may make movie gore more appealing.[23] If you don't feel the unpleasantness of disgust or respond empathically to the victims' agony, *anything* happening on the screen is okay.

It has already been established that gore watchers are more likely to be male, and that male psychopaths outnumber their female counterparts by two to one.[24] By contrast, women score higher on measures of empathy than men throughout life, and relative to overall brain size, women have proportionally larger insulas than men do.[25] Women are the more empathic and the more disgust-sensitive sex. At the same time, it is interesting to note that empathy can be used for malevolence. A torturer can use empathy to intuit how to increase the victim's suffering. In sports, business, and the social battlefield, empathy can be used to ascertain an opponent's weaknesses and then strategically implemented to manipulate and exploit one's rivals.[26] Maybe this is why women are more psychologically cruel than men—they know what to say and do to hurt the most. It has been well established that both sexes have equal potential for aggression, but that females typically enact their cruelty through psychological and emotional means while males are more physically violent.[27]

IS EMPATHY UNIQUELY HUMAN?

The emotional experience of disgust is unique to humans, and it develops in complexity as our brain and mental capacities mature and evolve. Given the interconnectedness between disgust and empathy, a logical question is whether empathy is uniquely human too. From what I have learned, the answer seems to be yes.

All female mammals from horses to humans, take care of their young. Therefore, the animals in our taxanomic class are predisposed to empathy because it is rooted in the necessity of offspring care. However, humans are more "immature" at birth and have the longest postpartum development of any mammal. A foal stands up and starts to walk within an hour after birth; our babies take about a year to do the same. Without a high degree of human maternal empathy, screaming infants would barely survive their first year. However, when a horse mother takes care of her distressed and hungry foal, it doesn't matter whether she *feels* what her foal is feeling; the behavior just requires that she knows what to do to make the foal stop mewling: feed it. Empathy goes well beyond responding appropriately to distress. Empathy requires having the psychological capacity for emotional concern, comfort, and care.

Research has shown that our primate cousins seem to show emotional concern and will console victims of aggression to whom they are unrelated. A bystander chimpanzee who sees one chimp beat up another will go over to the loser and gently put his arm around the defeated chimp's shoulders.[28] However, we can't know whether this consoling behavior implies that the "soothing" chimp is also feeling the hurt chimp's emotional and physical pain. Caring, though clearly related to empathy, does not require that you feel what the other is feeling; it just entails that you know the other is hurt and respond with nurturance. Likewise, being sympathetic doesn't require that you feel someone else's suffering; it just means that you act in a benevolent way when someone else is in distress. By contrast, empathy compels you to feel what I feel (even when you don't want to feel it), and when you are consoling me you feel my pain too.

Indeed, the major motivation for empathic consoling stems from wanting to decrease *our own* suffering. By soothing another's pain, we are vicariously relieved of their (and our) torment. Empathy is fundamentally selfish. I want you to feel better so that I can stop feeling lousy. Currently we have no way of knowing if the chimp who is consoling the loser of a fight is sharing the loser's pain, because they can't tell us, and their brains haven't been scanned while they are engaged in this behavior. For now, the only species which we can be certain shows self-involved, insula-based empathy is us.

We are also the only species who can feel empathy for victims whose suffering we do not see and whose lives are entirely removed from ours.[29] When people in the US send money to tsunami–earthquake victims in Japan, some proportion of these donors feel the pain and misery of people whom they have never met and have nothing in common with in either everyday life or specific crisis terms. Furthermore, the human capacity for language enables a higher form of empathy than any other animal can experience, and we can have conversations that almost sound ridiculous, such as "I understand that you understand that I understand your misery." Moreover, the empathy needed to make moral judgments about concepts such as the invasion of privacy imposed by airport body scanners when you haven't been on a plane in ten years is unique to humans. You need to empathically put yourself in the stocking feet of the traveler to know what it would feel like to have some stranger see the outline of your naked form in order to make a judgment about the rightness or wrongness of these machines. Therefore, although other mammals show nurturance, caring, and helping behaviors, the complexities of empathy, like the complexities of disgust, demand the uniquely intricate level of cognitive capacity that only we possess.

THE DISGUST AND EMPATHY BARRIER

Evangeline took care of the elderly, infirm, and nearly dead all day. She took them to the toilet, changed their diapers, bathed them, dressed their wounds—and she was easy, happy, and carefree with her charges and her work. But when her strong and able son was diagnosed with testicular cancer, she couldn't bear to be near him. His illness required surgery and one testicle had to be removed. When he was first recovering he needed someone to stay home with him, but Evangeline couldn't stand to see him, and she couldn't understand herself. Compared to what she dealt with every day, attending to him should be nothing—but it wasn't nothing. It was everything, and it was repulsive.

Why do some people feel more repulsion toward an intimate's injury than to a stranger's? I would feel much more disgusted if my husband's eyes were bloodily stitched up than if I saw the same injury on a stranger, male or female, because my husband is more connected to me than any stranger is. The level of intimacy and connection we have with someone parallels the level of disgust we can have toward their misfortunes. This is why it is relatively easy for us to disregard victims in foreign lands. For most of us, news that two hundred people were killed by a bomb in Iraq makes us feel upset, but the emotion tends to be fleeting and our thoughts turn to the next news story as it reaches our eyes and ears. But if two hundred people in the neighboring town were killed by a bomb, we would be beside ourselves and unable to think about anything else. This is in part because we may actually know some of them, but more importantly, it is because we can more easily put ourselves in their place and we realize that we could have been one of them. *We feel for ourselves.* Consider your reaction to 9/11. If you knew no one involved in the tragedy, you still likely felt much more distress about the victims there than about the many more thousands of people, including US soldiers, who have died in Iraq and Afghanistan since then. Distance makes the heart grow colder.

The greater the level of intimacy, the greater the level of empathy and disgust we feel toward someone. And because we are most intimate with ourselves, we feel the most disgust and the most empathy for ourselves. I haven't had my eyes cut or stitched, but when I was about twenty I got an extremely bad sunburn on my eyelids, though I wasn't aware of the severity until the next morning. When I woke up and couldn't open my eyes beyond a crack, I stumbled into the bathroom to find out why. When I looked at myself in the mirror and saw my horrifically puffed-up, red, and blistered eyelids I nearly fainted and threw up—the classic physiological reactions of disgust.

Though doctors and nurses need empathy in order to be caring practitioners, they are trained to regulate their emotional sensitivity so that they don't empathize too much. If doctors empathized the way most of us do, they would be too exhausted and upset to do their jobs well. Just how deep this professional distancing goes was shown in an experiment in which the brains of doctors and non-doctors were scanned while they watched a video of a hand being pricked by a needle.[30] The scans showed that the doctors' insulas were underactivated compared to those of non-doctors, but the doctors' brains showed comparatively greater activation in the areas of the brain that control reasoning and planning. In other words, experience and training enabled the doctors to down-regulate their empathy and think about the situation in pragmatic terms. For this same reason, doctors and nurses aren't much disgusted when a patient loses control of his bowels or comes into the emergency ripped open and gushing blood. The barrier to empathy keeps them from being overly disgusted. This is why Evangeline broke down—she couldn't distance herself from her son the way she can from the invalids she takes care of. Her empathy barrier failed her because she is too emotionally close to him. This is why surgeons aren't allowed to operate on family members, despite their training; their emotional involvement could lead to irrational decisions and mistakes.

TURN THAT GRIMACE UPSIDE DOWN

Disgust enables some of the worst interpersonal behavior. It can make us abandon valuable opportunities, commit grim social, personal, and moral mistakes, and lead to prejudice and murder. Disgust also feels ugly inside ourselves. When we are repulsed by the obese passenger in the airplane seat beside us, or feel loathing toward a stranger for her politics, we shudder a little at ourselves for the blackness that this feeling creates within us. And this feeling makes us feel ugly and disgusting too. For this reason alone, we might want to try not to be disgusted by people or things when the emotion serves no purpose. The next time you find yourself muttering, "That's disgusting," especially if it is about another person, ask yourself why you're feeling that way and if you really need to. Once you have dissected your emotions, you can decide whether they are worthy and worthwhile.

Reasoning can be our savior from disgust. Many things that disgust us have no logical reason to be avoided. In other words, they are not actually dangerous. Likewise, some things which we might be better served to be disgusted by don't bother us because our culture and our minds have made them okay. For example, cockroaches and garden snakes are much less likely to harm or sicken us than public computers or the exotic reptiles that are currently trendy pets. Thanks at least in part to charming insurance company ads, cute green geckos are one of the most popular pets in the US today, but geckos carry more strains of salmonella than any reptile ever studied. Approximately five hundred people die from salmonella in the US each year, and infants are especially susceptible to salmonella from reptiles (as opposed to salmonella from meat or eggs).[31] The number of salmonella-related illnesses is on a rapid rise in America and it closely tracks the increase in reptile pet sales.[32] Although the boost in cold-blooded house pets may indicate how far-reaching and effective creative marketing can be, it is foolhardy for us to be so easily beguiled by charismatic lizards.

The core of disgust, death, can also be better managed if we are willing to openly acknowledge and accept it rather than have it be the ever-hovering elephant in the corner. Mr. Greco, the funeral director I interviewed, lived with death professionally for over fifty years and wasn't disgusted by it or afraid of it. He had accepted and acknowledged death as a fact of life and was at peace. What helped him, he said, was that he dealt with death constantly. Just as we can inure ourselves to the revulsion of snakes, so can we reduce our abhorrence of death. I am not recommending loitering around mortuaries, rather simply practicing thinking about death in contemplative terms rather than denying it or being terrorized by it. This will make it easier to accept, and less debilitating to think about. And if we are less traumatized by death, there is good reason to believe that we will be kinder and more accepting toward others. It is our minds that make things disgusting or not to us, and we can actively use our minds to take the disgust out of them.

We can also actively monitor how our environment might be influencing us, and guard against being manipulated by dehumanizing propaganda or external manipulations of disgust. Listen carefully to the message that the hardliners are giving you. Does it make sense that another group of people is equivalent to vermin? Pay attention to your environment and what is in it. Is the courtroom unusually tidy or dirty? Were you just mailed a scented scarlet letter or flyer? Do you feel disgusted but don't know why? Is the person you're interacting with bothering you because of how they look, their sexual orientation, or their political beliefs? Slow down, think, and try to figure out why you are disgusted. Can you neutralize your feelings in order to be more just? Disgust can be crippling, but we can also walk away from it.

Empathy toward other people is broken when we have no understanding or connection to them. When a group is dehumanized and vilified as "cockroaches" or "rats," we respond to them as if they were vermin and do not think of them as human. Our disgust toward them becomes abstracted and depersonalized, and based on how they might contaminate us if we made physical contact with them. We even dehumanize people to the point of making them equivalent to disgusting inert objects. In a study conducted at Princeton University, the brains of participants were scanned while they looked at pictures of elderly people, people with disabilities, drug addicts, and homeless people.[33] When the participants saw pictures of the elderly and disabled, the brain area involved in social–people processing—the medial prefrontal cortex—was activated and so was the anterior insula. These people elicited disgust, but they also aroused pity and were still viewed as human. By contrast, when the college students looked at pictures of drug addicts and homeless people, there was no activity in the medial prefrontal cortex, and only the anterior insula was engaged. In fact, the students' brains lit up to the drug addicts and homeless in the same way as they did when they were shown pictures of vomit or an overflowing toilet. In other words, drug addicts and the homeless were viewed as subhuman and literally perceived as one would see a disgusting "thing."

This study not only shows how casually we dehumanize drug addicts and homeless people, but that we have internalized their "worthlessness." If we don't reflect on the fact that we so easily and dispassionately can relegate certain people to a subhuman realm, the potential for dangerous social consequences is very high. It is likely that during times of dehumanizing propaganda—when a particular group is analogized to cockroaches or rats—the "verminized" group would also elicit brain activity equivalent to seeing unflushed toilets, and not other human beings.

EMBRACING YOUR INNER DIRT

Disgust is not an evil emotion. Disgust is the by-product of being an empathetic and civilized person. You are not "bad" if you think that eating decomposing shark meat is disgusting, or for being repulsed by the scabby beggar. These things could in fact be accurate signals of disease, danger, and ultimately death. But not all of our disgusts are accurate threats. Being disgusted by immigrants, homosexuals, and drug addicts usually has no physically protective value. Yet it is because we are "civilized" and live within an integrated social network that we have learned to be, and are socially reinforced to be, disgusted by them. Nevertheless, we can be better than our erroneous disgusts.

Disgust is in our mind and fundamentally under our control. It is our psychology that makes cheddar cheese or casu marzu disgusting or delicious, homosexuality acceptable or not, scatological porn erotic or repulsive, foreigners innocuous or sickening, and our own farts or someone else's humorous or nauseating. We can decide which path to take, or at least ask ourselves why we are feeling the way we do and whether it is useful. We can decide to be disgusted or not.

Disgust is a highly socialized emotion. It depends on our being a part of an enmeshed social group, as well as understanding the importance of reciprocity within the social group. In the same way that we are disgusted by disease and should avoid it because it can corrupt our health, liars and cheaters are morally repugnant and should be avoided because they can corrupt our way of life. The abstract and egocentric aspects of disgust and empathy also underlie the fundamentals of politics. Politics involves understanding and manipulating the wants and goals of others and organizing group dynamics. Moral and social disgust—whom to avoid, exclude, or disapprove of—is a basic political tactic. "Vote for me and we'll keep *them* out and make *this* a law. Alongside this is the ability to exploit empathy and the skillful ways in which the politician seems to understand and take on our problems, and promises to make life better for all.

Disgust and empathy are about "me." Disgust, in particular, is about what unpleasantness could happen to us and our instinct to survive. How will this contaminate me? How can I get away from the ugliness? Empathy is egocentric but it also elicits nurturing: I want you to feel better so that I don't have to feel bad. At the same time that disgust and empathy are egocentric, they are also the cornerstones of human social life. They continually come into play, together and separately, in our interactions with other people. Therefore, rather than being disgusted and shunning a disfigured person or a group that has been dehumanized, can we try to alleviate what is disgusting us about them by being nurturing instead? If the man with the stitched-up eye had been alone and I had gone over to see if I could help him, would I have become less disgusted by him? I think so. The more we examine and reflect about the world we live in and the people around us, the more mindfully and compassionately we will live.

WHAT DISGUST IS

Disgust evolved from a simple mechanism that helped us avoid swallowing poison to one that warns us of death by the slow process of disease, and is ultimately about the uniquely human awareness of our fragile and finite mortality. Anything that triggers reminders of these issues, from slugs, to skeletons, to the shattering of our worldview, can elicit disgust in us. All disgusts motivate avoidance, but what the specific fleeting feelings of disgust are cannot be pinned down. Disgust can be obvious, visceral, and simplistic, or abstract, philosophical, and complicated.

Disgust is the most enigmatic of all emotions. Each of us is different when it comes to how disgustable we are, and what we personally find most horrific. Yet, while our sensitivity to disgust is a firm feature in each of our unique personalities, it can be changed by the whim of the moment. We can control disgust with our minds, and manipulate disgust in ourselves, and use it to influence others. Disgust comes in many forms and encompasses multiple states of mind and body from making us turn away from rotting food, to the full-body cringe that hearing about eyeball tattooing evokes, to the feeling you have about your neighbor's incest, to your opinion about the death penalty. It can be triggered by words that make us vomit, and conditioned linguistically to make our anger become repulsion. Disgust is sculpted by culture, but then it guides culture—and shapes our culinary practices, our lusts, our laws, and our customs.

The emotion of disgust is universal—every neurologically healthy person can experience it—but it is not innate. Disgust is the instinct that has to be learned. Disgust depends on a healthy and mature brain, not simply because of the required working hardware but because of the necessity for self-aware, elaborate thought. Ultimately, disgust is about rejection—and at its heart the rejection of our deepest fear, our inevitable death.

It is in this impulse of rejection that it becomes revealed how our human experience of disgust is a luxury. Only when we have abundance, opportunity, and security can we afford to be revolted by rotted food, an ugly potential mate, or a neighbor's licentious behavior. When survival is the only thing that counts we will eat anything we can, mate with whomever we can, and enlist the help of anyone who is available. Other animals don't have the luxury to choose between myriad options for sustenance and companionship because death is so much more at hand. It is a sign of our good fortune that we have the privilege to be disgusted. The victim of a subzero airplane crash will grasp at anything for food and huddle next to the person beside him for warmth no matter how pockmarked or dirty they are. When the separation between our life and our death is wavering in front of us, we will take our chances with survival rather than recoil in disgust and meet certain annihilation. Disgust means that we have the luxury to choose between what we would prefer to have and what we must have now in order to survive to the next moment.

From our ability to understand the privilege of abundance to the inevitability of our death, disgust holds a mirror up to us. This is why, by unraveling disgust, we can more fully understand what it means to be human and our personal, unique version of it.

Acknowledgments

I owe most of my thinking, opinions, and ideas to the many disgust experts whose books and articles I read over the last several years. I am particularly thankful for the inspiration and insights of (in alphabetical order): Ernest Becker, Valerie Curtis, Mary Douglas, Daniel Fessler, Jamie Goldenberg, Jeff Greenberg, Jonathan Haidt, Susan Miller, William Miller, Martha Nussbaum, Tom Pyszczynski, Paul Rozin, Mark Schaller, Jeff Schimel, Sheldon Solomon, Richard Stevenson, and all of their research collaborators. Most of all I am indebted to Paul Rozin, who first enticed me with disgust many years ago and who has been my friend and mentor on the topic ever since. Kind colleagues whom I pestered and who generously offered me wisdom and assistance were: Jonathan Haidt, Reiner Sprengelmeyer, Graham Davey, Jean Decety, Jean-Charles Chabat, Victoria Esses, and Pamela Dalton.

Special appreciation goes to my efficacious and ever-supportive agent, Wendy Strothman, as well as the many people at W. W. Norton who helped make this book happen, in particular my very insightful and talented editor, Angela von der Lippe, and her assistant, Laura Romain.

I am indebted to my friends and family members who told me stories, gave me ideas, and alerted me to pertinent facts and findings, especially: Judith Herz, Jamie Poy, Nathaniel Herz, Kathleen McCann,

John McCann, Akiko Yamamoto, Theresa White, Rachel Tyndale, Liz Holstein, Amy Hagan, and Tanya Anstey. Additional gratitude goes to Theresa White for our penetrating conversations on empathy and the meaning of moral and social disgust, and to the Brown University students who participated in my 2010 and 2011 course on "The Psychology of Aversion" for spirited and idea-generating discussions. Notable thanks also go to Robert and Susan Iannotti for enlightening conversations and to Joy Robinson, who invited me to judge the National Rotten Sneaker Contest, from which the idea for this book was spawned.

My deepest appreciation goes to those who warmed my heart and mind at home while I worked on this book: my dear springer spaniel, Molly, for whom rabbit poop is caviar and who granted me lots of opportunities for "thinking walks"; my mother, for expert English advice and always being available at the other end of the telephone; and most of all, my wonderful husband, Jamie, who gave me many great ideas, provided constant support, and was my perpetual sounding board.

Notes

PREFACE

1 Paul Rozin quoted in J. Gorman, "A Perk of Our Evolution: Pleasure
  in Pain of Chilies," *New York Times*, September 20, 2010, http://www
  .nytimes.com/2010/09/21/science/21peppers.html?_r-1&emc-etal.

CHAPTER 1: LET'S EAT

1 http://www.meguminatto.com/history.html.
2 http://blog.hotelclub.com/top-10-stinky-cheeses/.
3 J. Brumberg-Kraus and B. Dexter Dyer, "Cultures and Cultures: Fermented Foods as Culinary Shibboleths," in *Cured, Fermented, and Smoked
  Foods: Proceedings of the Oxford Symposium on Food and Cookery 2010*
  (Totnes, UK: Prospect Books, 2011), 56–65.
4 *No Reservations*, Travel Channel, season 1, episode 2, August 1, 2005.
5 Brumberg-Kraus and Dexter Dyer, "Cultures and Cultures."
6 D. Kraemer, *Jewish Eating and Identity Through the Ages* (London: Routledge, 2009), 68.
7 Brumberg-Kraus and Dexter Dyer, "Cultures and Cultures."
8 Andrew Zimmern in his episode on Bolivia, *Bizarre Foods*, Travel Channel, April 1, 2008.
9 R. Wetzer and N. D. Pentcheff, "Insects are crustaceans," Natural History Museum of Los Angeles County, 2002; J. L. Boore, D. V. Lavrov,

and W. M. Brown, "Gene translocation links insects and crustaceans," *Nature* 392 (1998), 667–68.

10 http://maineslobster.com/history-of-maine-lobster-fishing.

11 http://www.lobsters.org/tlcbio/biology.html.

12 P. Guthrie, Cox News Service, July 7, 1999.

13 P. Apari and L. Rózsa, "Deal in the womb: fetal opiates, parent–offspring conflict, and the future of midwifery," *Medical Hypotheses* 67 (2006), 1189–94.

14 Material for this section from N. Parrado, *Miracle in the Andes* (New York: Three Rivers Press, 2006).

15 Information gathered from *Eating with Cannibals*, National Geographic Channel, April 9, 2011.

16 For details on odor preferences and odor learning see R. Herz, *The Scent of Desire: Discovering Our Enigmatic Sense of Smell* (New York: Harper-Collins, 2007).

17 "In Iowa, Indiana state fairs, weird foods reign," NPR, August 20, 2010; listener responses read August 24, 2010.

18 P. Rozin, M. Markwith, and C. Stoess, "Moralization and becoming a vegetarian: the transformation of preferences into values and the recruit-ment of disgust," *Psychological Science* 8 (1997), 67–73.

19 R. I. Stein and C. J. Nemeroff, "Moral overtones of food: judgments of others based on what they eat," *Personality and Social Psychology Bulletin* 21 (1995), 480–90.

20 Buffalo or bull testicles, flattened and deep-fried.

21 Rozin, Markwith, and Stoess, "Moralization."

22 D. M. Fessler, A. P. Arguello, J. M. Mekdara, and R. Macias, "Disgust sensitivity and meat consumption: a test of an emotivist account of moral vegetarianism," *Appetite* 41 (2003), 31–41.

23 P. Rozin, "Moralization," in A. Brandt & P. Rozin, eds., *Morality and Health* (New York: Routledge, 1997).

24 P. Rozin and L. Singh, "The moralization of cigarette smoking in the United States," *Journal of Consumer Behavior* 8 (1999), 321–37.

25 J. Horsman and J. Flowers, *Please Don't Eat the Animals* (Sanger, CA: Quill Driver Books, 2007).

26 *World Population Prospects, 2008 revision*. United Nations, 2009.

27 As seen on *Bizarre Foods*, Travel Channel, season 1, episode 11, August 6, 2007.

28 The Explorers Club is an international multidisciplinary professional society dedicated to the advancement of field research and the ideal that it is vital to preserve the instinct to explore. Its members have included Sir Edmund Hillary and Neil Armstrong.

29 Personal communication with Explorers Club members Edward Lovett, June 4, 2011, and Gene Rurka, June 5, 2011.

30 J. Ramos-Elorduy, *Creepy Crawly Cuisine: The Gourmet Guide to Edible Insects* (Rochester, VT: Park Street Press, 1998). D. G. Gordon, *The Eat a Bug Cookbook* (Berkeley, CA: Ten Speed Press, 1998).

31 Y. P. Zverev, "Effects of caloric deprivation and satiety on sensitivity of the gustatory system," *BMC Neuroscience* 5:5 (2004).

32 D. M. Small et al., "Changes in brain activity related to eating chocolate: from pleasure to aversion," *Brain* 124 (2001), 1720–33.

33 Record set in 2007.

34 P. Rozin, J. Haidt, and C. R. McCauley, "Disgust," in M. Lewis, J. M. Haviland-Jones, and L. F. Barrett, eds., *Handbook of Emotions*, 3d ed (New York: Guilford Press, 2008), 757–76.

35 P. Rozin and D. Schiller, "The nature and acquisition of a preference for chili pepper by humans," *Motivation and Emotion* 4 (1980), 77–101.

CHAPTER 2: A SNAPSHOT OF DISGUST

1 P. Ekman and W. Friesen, "Constants across cultures in the face and emotion," *Journal of Personality and Social Psychology* 17 (1971), 124–29.

2 D. Galati, K. R. Scherer, and P. E. Ricci-Bitti, "Voluntary facial expression of emotion: comparing congenitally blind with normally sighted encoders," *Journal of Personality and Social Psychology* 73 (1997), 1365–79.

3 J. M. Susskind et al., "Expressing fear enhances sensory acquisition," *Nature Neuroscience* 11 (2008), 843–50.

4 Alkaloids have a basic pH (pH above 8).

5 This is the DS–R, "Disgust Scale (Revised)." It was originally published in J. Haidt, C. McCauley, and P. Rozin, "Individual differences in sensitivity to disgust: A scale sampling seven domains of disgust elicitors," *Personality and Individual Differences* 16 (1994), 701–13. Modifications resulting in the revised version were suggested by B. O. Olatunji et al., "The disgust scale: item analysis, factor structure, and suggestions for refinement," *Psychological Assessment* 19 (2007), 281–97. For more information see http://people.virginia.edu/~jdh6n/disgustscale.html.

6 The questionnaire reproduced here is a 27-item revised version of the original 32-item Disgust Scale, published in: Haidt, McCauley, and Rozin, "Individual differences in sensitivity to disgust." Both versions of the questionnaire have been widely administered, though the revised version is used more now.

7 In a few cases in this book other aspects of disgust have been tested and will be discussed.

8 J. Haidt, P. Rozin, C. McCauley, and S. Imada, "Body, psyche, and culture: the relationship between disgust and morality," *Psychology and Developing Societies* 9 (1997), 107–31; D. M. Fessler, A. P. Arguello, J. M. Mekdara, and R. Macias, "Disgust sensitivity and meat consumption"; V. Curtis, R. Aunger, and T. Rabie, "Evidence that disgust evolved to protect from risk of disease," *Proceedings of the Royal Society B* 271 (2004), S131–33.

9 M. Oaten, R. J. Stevenson, and T. I. Case, "Disgust as a disease avoidance mechanism," *Psychological Bulletin* 135 (2009), 303–21.

10 J. Haidt, C. McCauley, and P. Rozin, "Individual differences in sensitivity to disgust: A scale sampling seven domains of disgust elicitors," *Personality and Individual Differences* 16 (1994), 701–13.

11 A. Aleman and M. Swart, "Sex differences in neural activation to facial expressions denoting contempt and disgust," *PLOS One* 3 (2008), 1–7.

12 S. C. Roberts et al., "Manipulation of body odour alters men's self-confidence and judgments of their visual attractiveness by women," *International Journal of Cosmetic Science* 31 (2009), 47–54.

13 C. Zahn-Waxler, M. Radke-Yarrow, E. Wagner, and M. Chapman, "Development of concern for others," *Developmental Psychology* 28 (1992), 126–36.

14 C. Classen, D. Howes, and A. Synnott, *Aroma: The Cultural History of Smell* (New York: Routledge, 1994).

15 J. M. Cisler, B. O. Olantunji, and J. M. Lohr, "Disgust, fear, and the anxiety disorders: a critical review," *Clinical Psychology Review* 29 (2009), 34–46.

16 A. C. Page, "Blood-injury injection fears in medical practice," *Medical Journal of Australia* 164 (1996), 189. B. Deacon and J. Abramowitz, "Fear of needles and vasovagal reactions among phlebotomy patients," *Journal of Anxiety Disorders* 20 (2006), 946–60.

17 Deacon and Abramowitz, "Fear of needles."

18 M. Muris et al., "The effects of verbal disgust- and threat-related information about novel animals on disgust and fear beliefs and avoidance in children," *Journal of Clinical Child and Adolescent Psychology* 38 (2009), 551–63.

19 J. A. J. Smits, M. J. Telch, and P. K. Randall, "An examination of decline in fear and disgust during exposure based treatment," *Behaviour Research and Therapy* 40 (2002), 1242–53.

20 Pamela Dalton, personal communication, October 21, 2010.

21 Classen, Howes, and Synnott, *Aroma.*

22 D. L. Mosher, "Sex differences, sex experience, sex guilt and explicit sexual films," *Journal of Social Issues* 29 (1973), 95–112.

23 D. M. T. Fessler and C. D. Navarette, "Third-party attitudes toward sibling incest evidence for Westermarck's hypothesis," *Evolution and Human Behavior* 25 (2004), 277–94.

24 Y. Inbar, D. A. Pizarro, and P. Bloom, "Conservatives are more easily disgusted than liberals," *Cognition and Emotion* 23 (2009), 714–25. J. Haidt and J. Graham, "When morality opposes justice: conservatives have moral intuitions that liberals may not recognize," *Social Justice Research* 20 (2007), 98–116.

25 S. C. Widen and J. A. Russell, "The 'disgust face' conveys anger to children," *Emotion* 10 (2010), 455–66.

26 H. Aviezer et al., "Angry, disgusted or afraid?" *Psychological Science* 19 (2008), 724–32.

27 See S. Tracy et al., "Enteroviruses, type 1 diabetes and hygiene: a complex relationship," *Reviews in Medical Virology* 20 (2010), 106–16, and http://www.suite101.com/content/early-germ-exposure-boosts-immune-function-a76488.

28 J. Russell, "Children's recognition of emotion from faces correlates with learning emotion words," Annual Meeting of the Association for Psychological Science, Boston, 2010.

29 W. I. Miller, *The Anatomy of Disgust* (Cambridge, MA: Harvard University Press, 1997).

30 Elizabeth Holstein, personal communication, January 15, 2010.

31 L. Malson, *Wolf Children*, translated by E. Fawcett, P. Ayrton, and J. White (New York: Monthly Review Press, 1972). (Originally published in French, 1964.)

32 J. Itard, *The Wild Boy of Aveyron* (New York: Meredith Company, 1962), 100. (Originally published in 1801.)

33 S. Chevalier-Skolnikoff, "Facial expression of emotion in nonhuman primates," in P. Ekman, ed., *Darwin and Facial Expression: A Century of Research in Review* (New York: Academic Press, 1973), 11–90.

34 http://news.directory.com/human/biologist-enthralls-kids-with-maggot-art.html#mkcpgn=rssnws1.

35 S. B. Miller, *Disgust: The Gatekeeper Emotion* (Hillsdale, NJ: Analytic Press, 2004).

36 R. J. Stevenson and B. M. Repacholi, "Does the source of an interpersonal odour affect disgust? A disease risk model and its alternative," *European Journal of Social Psychology* 35 (2005), 375–401.

37 T. I. Case, B. M. Repacholi, and R. J. Stevenson, "My baby doesn't smell as bad as yours: the plasticity of disgust," *Evolution and Human Behavior* 27 (2006), 357–65.

38 Stevenson and Repacholi, "Does the source of an interpersonal odour . . ."

39 R. Buchsbaum, *Animals Without Backbones*, 2d ed, (Chicago: University of Chicago Press, 1948).

40 http://www.davesdaily.com/funnynews/pickingnose_07-05.htm.

41 http://www.wacktrap.com/legal/laws/nose-picking-driving-beats-wireless-device-law.

42 For more information on the psychology of smell see R. Herz, *The Scent of Desire: Discovering Our Enigmatic Sense of Smell* (New York: Harper-Collins, 2007).

43 R. S. Herz, and J. von Clef, "The influence of verbal labeling on the perception of odors: Evidence for olfactory illusions?" *Perception* 30 (2001), 381–91.

CHAPTER 3: DISGUST ON THE BRAIN

1 George Huntington, MD, "On Chorea," *The Medical and Surgical Reporter: A Weekly Journal* (Philadelphia: S. W. Butler) 26, no. 15 (April 13, 1872), 317–21.

2 SM also had some difficulty recognizing anger and surprise faces.

3 R. Adolphs, D. Tranel, H. Damasio, and A. Damasio, "Impaired recognition of emotion in facial expressions following bilateral damage to the human amygdala," *Nature* 372 (1994), 669–72.

4 The inability to experience disgust is not a symptom of all HD patients.

5 C. J. Hayes, R. J. Stevenson, and M. Coltheart, "Disgust and Huntington's disease," *Neuropsychologia* 45 (2007), 1135–51.

6 R. Sprengelmeyer, U. Schroeder, A. W. Young, and J. T. Epplen, "Disgust in pre-clinical Huntington's disease: a longitudinal study," *Neuropsychologia* 44 (2006), 518–33.

7 J. Wynbrandt and M. D. Ludman, *The Encyclopedia of Genetic Disorders and Birth Defects* (New York: Facts on File, 1991).

8 The structures of the basal ganglia are the striatum, pallidum, substantia nigra, and subthalamic nucleus.

9 S. K. Kamboj and H. V. Curran, "Scopolamine induces impairment in the recognition of human facial expressions of anger and disgust," *Psychopharmacology* 185 (2006), 529–35.

10 The temporal and frontal lobes are involved in higher cognitive func-

tions. The parietal lobe controls our sense of where we are in the world spatially and by touch.

11 B. Wicker et al., "Both of us disgusted in my insula: the common neural basis of seeing and feeling disgust," *Neuron* 40 (2003), 655–64; P. Wright et al., "Disgust and the insula: fMRI responses to pictures of mutilation and contamination," *Neuroreport* 15 (2004), 2347–51; M. L. Phillips et al., "A specific neural substrate for perceiving facial expressions of disgust," *Nature* 389 (1997), 495–98.

12 M. J. Thieben et al., "The distribution of structural neuropathology in pre-clinical Huntington's disease," *Brain* 125 (2002), 1815–28.

13 R. Sprengelmeyer et al., "Facial expression recognition in people with medicated and unmediated Parkinson's disease," *Neuropsychologia* 41 (2003), 1047–57; M. D. Pell and C. L. Leonard, "Facial expression decoding in early Parkinson's disease," *Cognitive Brain Research* 23 (2005), 327–40; K. Wang et al., "Impairment of recognition of disgust in Chinese with Huntington's or Wilson's disease," *Neurophychologia* 41 (2003), 527–37.

14 W. Penfield and M. E. Faulk, "The insula: further observations on its function," *Brain* 78 (1955), 445–70.

15 M. L. Phillips et al., "A specific neural substrate for perceiving facial expressions of disgust," *Nature* 389 (1997), 495–98.

16 http://www.disabled-world.com/artman/publish/famous-ocd.shtml.

17 R. Sprengelmeyer et al., "Disgust implicated in obsessive-compulsive disorder," *Proceedings of the Royal Society B*, 264 (1997), 1767–73.

18 K. M. Corcoran, S. R. Woody, and D. F. Tolin, "Recognition of facial expression in obsessive-compulsive disorder," *Journal of Anxiety Disorders* 22 (2008), 56–66.

19 D. S. Husted, N. A. Shapira, and W. K. Goodman, "The neurocircuitry of obsessive-compulsive disorder and disgust," *Progress in Neuro-Psychopharmacology and Biological Psychiatry* 30 (2006), 389–99.

20 D. J. Stein et al., "Neurocircuitry of disgust and anxiety in obsessive-compulsive disorder: a positron emission tomography study," *Metabolic Brain Disorders* 21 (2006), 267–77.

21 See J.-Y. Rotge et al., "Provocation of obsessive-compulsive symptoms: a quantitative voxel-based meta-analysis of functional neuroimaging studies," *Journal of Psychiatry and Neuroscience* 33 (2008), 405–12.

22 H. Hermesh et al., "Orbitofrontal cortex dysfunction in obsessive compulsive disorder II: olfactory quality discrimination in OCD," *European Neuropsychopharmacology* 9 (1999), 415–20.

23 GABA, or gamma-Aminobutyric acid, is the primary inhibitory neurotransmitter in the mammalian brain. Neurotransmitters are endogenous

brain chemicals that modulate various functions and brain structures and are classified as either excitatory or inhibitory. Other neurotransmitters include serotonin, dopamine, norepinephrine, glutamate, and acetylcholine.

24 G. Zai et al., "Evidence for the gamma-amino-butyric acid type B receptor 1 (*GABBR1*) gene as a susceptibility factor in obsessive-compulsive disorder," *American Journal of Medical Genetics*, Part B, 134B (2005), 25–29.

25 G. Skoog and I. Skoog, "A 40-year follow-up of patients with obsessive-compulsive disorder," *Archives of General Psychiatry* 56 (1999), 121–27.

26 L. Martin et al., "Enhanced recognition of facial expression of disgust in opiate users receiving maintenance treatment," *Addiction* 101 (2006), 1598–1605.

27 Sprengelmeyer et al., "Disgust implicated in obsessive-compulsive disorder."

28 The National Center for Scientific Research is a collective of government-based research institutes in various locations throughout France.

29 Wicker et al., "Both of us disgusted."

30 I. L. Bernstein, "Learned taste aversions in children receiving chemotherapy," *Science* 200 (1978), 1302–03.

31 Lionel Dahmer, *A Father's Story* (New York: William Morrow, 1994), 28.

32 D. S. Kosson, Y. Suchy, A. R. Mayer, and J. Libby, "Facial affect recognition in criminal psychopaths," *Emotion* 2 (2002), 398–411.

33 R. E. Jack et al., "Cultural confusions show that facial expressions are not universal," *Current Biology* 19 (2009), 1543–48.

CHAPTER 4: GERM WARFARE

1 N. D. Wolfe, C. P. Donavan, and J. Diamond, "Origins of major human infectious diseases," *Nature* 447 (2007), 279–83.

2 http://www.avert.org/worldstats.htm; http://www.cdc.gov/hiv/resources/factsheets/us.htm.

3 K. Cambra, "Snakes on a plane," *Brown Medicine* 16 (2010), 28–35; K. E. Jones et al., "Global trends in emerging infectious diseases," *Nature* 451 (2008), 990–93.

4 http://www.cdc.gov/flu/about/disease/us_flu-related_deaths.htm.

5 A. Öhman and S. Mineka, "Fears, phobias and preparedness: Toward an evolved module of fear and fear learning," *Psychological Review* 108 (2011), 483–522.

6 A. K. Anderson et al., "Neural correlates of the automatic processing of threat facial signals," *Journal of Neuroscience* 23 (2003), 5627–33.

7 G. C. L. Davey, "The 'disgusting' spider: the role of disease and illness in the perpetuation of fear of spiders," *Society and Animals* 2 (1994), 17–25.

8 W. H. McNeil, *Plagues and People* (London: Penguin, 1976).

9 C. S. Crandall and D. Moriarty, "Physical illness stigma and social rejection," *British Journal of Social Psychology* 34 (1995), 67–83.

10 http://checkorphan.getreelhealth.com/grid/news/treatment/leper-colonies-on-the-road-to-extinction.

11 Life expectancy with birth year between 2010 and 2015 from *United Nations World Population Prospects 2008 Revision*, see http://esa.un.org/unpp/.

12 I. G. A. Bradshaw, *Elephants on the edge* (New Haven, CT: Yale University Press, 2009).

13 I. G. A. Bradshaw, "Not by bread alone: symbolic loss, trauma, and recovery in elephant communities," *Society and Animals* 12 (2004), 143–58.

14 M. Rubio-Godoy, R. Aunger, and V. Curtis, "Serotonin: a link between disgust and immunity?" *Medical Hypotheses* 68 (2007), 61–66.

15 M. Oaten, R. J. Stevenson, and T. I. Case, "Disgust as a disease avoidance mechanism," *Psychological Bulletin* 135 (2009), 303–21.

16 M. Schaller et al., "Mere visual perception of other people's disease symptoms facilitates a more aggressive immune response," *Psychological Science* 21 (2010), 649–52.

17 R. J. Stevenson, T. I. Case, and M. J. Oaten, "Frequency and recency of infection and their relationship with disgust and contamination sensitivity," *Evolution and Human Behavior* 30 (2009), 363–68.

18 The Big Five personality factors have been found to characterize all human personality. Determined by extensive, comprehensive, data-driven personality analysis, they are: openness to experience, conscientiousness, extroversion, agreeableness, and neuroticism. Conscientiousness is defined as a tendency to show self-discipline, act dutifully, and aim for achievement; to be planned rather than engage in spontaneous behavior.

19 R. R. McCrae, "NEO-PI-R data from 36 cultures," in R. R. McCrae and J. Allik, eds., *The Five-Factor Model of Personality Across Cultures* (New York: Kluwer Academic/Plenum, 2002).

20 H. S. Friedman and L. R. Martin, *The Longevity Project* (New York: Hudson Street Press, 2011).

21 G. C. L. Davey et al., "Familial resemblances in disgust sensitivity and animal phobias," *Behaviour Research and Therapy* 31 (1993), 41–50.

22 M. Schaller and D. R. Murray, "Pathogens, personality, and culture: disease prevalence predicts worldwide variability in sociosexuality, extraversion and openness to experience," *Journal of Personality and Social Psychology* 95 (2008), 212–21.

23 WHO data as of 2008. See http://www.nationmaster.com/graph/hea_int_dis_dea_rat-health-intestinal-diseases-death-rate.

24 D. P. Schmitt, "Sociosexuality from Argentina to Zimbabwe: a 48-nation study of sex, culture, and strategies of human mating," *Behavioral and Brain Sciences* 28 (2005), 247–311.

25 Schaller and Murray, "Pathogens, personality, and culture."

26 M. Laska et al., "Failure to demonstrate systematic changes in olfactory perception in the course of pregnancy: a longitudinal study," *Chemical Senses* 21 (1996), 567–71; E. L. Cameron, "Measures of human olfactory perception during pregnancy," *Chemical Senses* 32 (2007), 775–82.

27 C. D. Navarette, D. M. T. Fessler, and S. J. Eng, "Elevated ethnocentrism in the first trimester of pregnancy," *Evolution and Human Behavior* 28 (2007), 60–65.

28 P. Dalton, "Odor perception and beliefs about risk," *Chemical Senses* 21 (1996), 447–58.

29 Ibid.

30 A. P. R. Wilson et al., "Computer keyboards and the spread of MRSA," *Journal of Hospital Infection* 62 (2006), 390–92.

31 "Germs, Germs, Everywhere! Study of public surfaces produces surprising results," ATM Services, Inc., June 2006.

32 "What's Dirtier, Cell Phone or Toilet Seat?" *ABC News*, August 4, 2006.

33 N. Fierer et al., "Forensic identification using skin bacterial communities," *Proceedings of the National Academy of Sciences, USA* 107 (2010), 6477–81.

34 "Paper Money Breeds Disease," *New York Times*, March 15, 1904.

35 Dow Jones News, 1998.

36 T. M. Pope et al., "Bacterial contamination of paper currency," *Southern Medical Journal* 95 (2002), 1408–10.

37 C. J. Uneke and O. Ogbu, "Potential for parasite and bacteria transmission by paper currency in Nigeria," *Journal of Environmental Health* 69 (2007), 54–60.

38 M. Moss, "The Burger that Shattered her Life," *New York Times*, October 3, 2009.

39 Ibid.

40 Estimate from the Centers for Disease Control and Prevention, see http://digestive.niddk.nih.gov/ddiseases/pubs/bacteria/.

41 The Center for Science in the Public Interest compiled data from state and federal government reports, scientific articles, and news reports that focus on foods overseen by the Food and Drug Administration; meats are regulated by the US Department of Agriculture.

42 J. E. LeDoux, *The Emotional Brain* (New York: Simon & Schuster, 1996).

43 When the human brain reached the cognitive capacity that it is at now. Note, however, that physical and cultural evolution is a constant process and therefore our brains are likely slightly different than they were 50,000 years ago.

44 See V. Curtis and A. Biran, "Dirt, disgust and disease," *Perspectives in Biology and Medicine* 44 (2001), 17–31.

45 http://science.nationalgeographic.com/science/health-and-human-body/human-diseases/raging-malaria.html.

46 M. G. Rooks and W. S. Garrett, "Bacteria, food, and cancer," *F1000 Biology Reports* 3:12 (2011), doi: 10.3410/B3-12.

47 C. Zimmer, "How Microbes Define and Defend Us," *New York Times*, July 12, 2010.

48 D. M. T. Fessler and C. D. Navarette, "Meat is good to taboo," *Journal of Cognition and Culture* 3 (2003), 1–40.

49 Navarette, Fessler, and Eng, "Elevated ethnocentrism in the first trimester of pregnancy."

CHAPTER 5: DISGUST IS OTHER PEOPLE

1 Curtis, Aunger, and Rabie, "Evidence that disgust evolved to protect from risk of disease."

2 Miller, *The Anatomy of Disgust*.

3 Miller, *Disgust: The Gatekeeper Emotion*.

4 Curtis, Aunger, and Rabie, "Evidence that disgust evolved to protect from risk of disease."

5 "Findings," *Harper's*, August 2009, 84.

6 Examples are M. Mauss, *A General Theory of Magic* (1902), translated by R. Brian (New York: W. W. Norton, 1972); J. G. Frazer, *The Golden Bough: A Study in Magic and Religion* (1890), edited by T. G. Gaster (New York: Macmillan, 1959).

7 A. Meigs, *Food, Sex and Pollution: A New Guinea Religion* (New Brunswick, NJ: Rutgers University Press, 1984).

8 P. Rozin, L. Millman, and C. Nemeroff, "Sympathetic magic in disgust and other domains," *Journal of Personality and Social Psychology* 50 (1986),

703–12. C. Nemeroff and P. Rozin, "The contagion concept in adult thinking the United States: transmission of germs and interpersonal influence," *Ethos* 22 (1994), 158–86.

9 Oaten, Stevenson, and Case, "Disgust as a disease avoidance mechanism."

10 P. Rozin, M. Markwith, and C. McCauley, "Sensitivity to indirect contacts with other persons: AIDS aversion as a composite of aversion to strangers, infection, moral taint, and misfortune," *Journal of Abnormal Psychology* 103 (1984), 495–504.

11 A. C. Morales and G. J. Fitzsimmons, "Product contagion: changing consumer evaluations through physical contact with 'disgusting' products," *Journal of Marketing Research* 44 (2007), 272–83.

12 Nemeroff and Rozin, "The contagion concept."

13 J. A. Argo, D. W. Dahl, and A. C. Morales, "Positive consumer contagion: responses to attractive others in a retail context," *Journal of Marketing Research* 45 (2008), 690–701.

14 M. C. Nussbaum, *Hiding From Humanity: Disgust, Shame and the Law* (Princeton, NJ: Princeton University Press, 2004).

15 R. N. Proctor, *The Nazi War on Cancer* (Princeton, NJ: Princeton University Press, 1999), 46–48.

16 M. Addington, "The (thermo) dynamic body," *Biosense: The Future of Sweat*, symposium, New York, May 29, 2008.

17 R. A. Meckel, "Open-air schools and the tuberculosis child in early 20th century America," *Archives of Pediatric and Adolescent Medicine* 150 (1996), 91–96.

18 V. M. Esses, "The dehumanization of immigrants and refugees," Annual Meeting of the Association for Psychological Science, Boston, 2010; H. Markel and A. M. Stern, "The foreignness of germs: the persistent association of immigrants and disease in American society," *Millbank Quarterly* 80 (2002), 757–88.

19 The formula to calculate your body mass index (BMI) = (Your Weight in Pounds / Your Height in inches x Your Height in inches) x 703.

20 M. Tiggemann and E. D. Rothblum, "Gender differences in social consequences of perceived overweight in the United States and Australia," *Sex Roles* 18 (1988), 75–86.

21 C. S. Crandall, "Prejudice against fat people: ideology and self interest," *Journal of Personality and Social Psychology* 66 (1994), 882–94.

22 "F as in for Fat," Trust for America's Health, Executive report, July 2009, Robert Wood Johnson Foundation, www.healthyamericans.org.

23 http://www.cdc.gov/nchs/fastats/overwt.htm.

24 Y. Klimentidis et al., "Canaries in the coal mine: across-species analysis

of the plurality of obesity epidemics," *Proceedings of the Royal Society B* (2010), doi:10.1098/rspb.2010.1890.

25 R. Puhl, K. E. Henderson, and K. D. Brownell, "Social consequences of obesity," in P. Kopelman, I. Caterson, & W. Dietz, eds., *Clinical Obesity and Related Metabolic Disease in Adults and Children* (Oxford: Blackwell, 2005), 29–45.

26 M. V. Roehling, P. V. Roehling, and S. Pichler, "The relationship between body weight and perceived weight-related employment discrimination: the role of sex and race," *Journal of Vocational Behavior* 71 (2009), 300–18.

27 M. R. Hebl and L. M. Mannix, "The weight of obesity in evaluating others: a mere proximity effect," *Personality and Social Psychology Bulletin* 29 (2003), 28–38.

28 T. Parker-Pope, "An Older Generation Falls Prey to Eating Disorders," *New York Times*, March 29, 2011.

29 A small percentage of bulimics use fasting and excessive exercise, rather than purging, as a means to counteract their food binges. A variant called binge eating disorder is when out-of-control eating occurs along with self-loathing and depression, but there is no purge. These people are typically overweight.

30 N. Troop and A. Baker, "Food, body, and soul: the role of disgust in eating disorders," in B. O. Olatunji and D. McKay, eds., *Disgust and Its Disorders: Theory, Assessment, and Treatment Implications* (Washington, DC: American Psychological Association, 2009).

31 M. Boskind-White and W. C. White, *Bulimia/Anorexia: the Binge/Purge Cycle and Self-Starvation* (New York: W. W. Norton, 2001).

32 G. C. L. Davey, G. Buckland, B. Tantow, and R. Dallos, "Disgust and eating disorders," *European Eating Disorders Review* 6 (1998), 201–11; N. Troop, J. Treasure, and L. Serpell, "A further exploration of disgust in eating disorders," *European Eating Disorders Review* 10 (2002), 218–26.

33 B. Mayer et al., "Does disgust enhance eating disorder symptoms," *Eating Behaviors* 9 (2008), 124–27.

34 J. H. Park, M. Schaller, and C. S. Crandall, "Pathogen-avoidance mechanisms and the stigmatization of obese people," *Evolution and Human Behavior* 28 (2007), 410–14.

35 Project Implicit, an international Web-based research program run through Harvard University.

36 Park, Schaller, and Crandall, "Pathogen-avoidance mechanisms."

37 Ibid.

38 M. Schaller and L. A. Duncan, "The Behavioral Immune System: Its Evolution and Social Psychological Implications," in J. P. Forgas,

M. G. Haselton and W. von Hippel, eds., *Evolution and the Social Mind* (New York: Psychology Press, 2007), 293–307.

39 J. H. Park, J. Faulkner, and M. Schaller, "Evolved disease avoidance processes and contemporary anti-social behavior: prejudicial attitudes and avoidance of people with physical disability," *Journal of Nonverbal Behavior* 27 (2003), 65–87; C. R. Mortensen et al., "Infection breeds reticence: the effects of disease salience and self-perception and personality and behavior avoidance tendencies," *Psychological Science* 21 (2010), 440–47.

40 J. Faulkner, M. Schaller, J. H. Park, and L. A. Duncan, "Evolved disease-avoidance mechanics and contemporary xenophobic attitudes," *Group Processes and Intergroup Relations* 7 (2004), 333–53; C. D. Navarette and D. M. T. Fessler, "Disease avoidance and ethnocentrism: the effects of disease vulnerability and disgust sensitivity on intergroup attitudes," *Evolution and Human Behavior* 27 (2006), 270–82.

41 http://penelope.uchicago.edu/~grout/encyclopaedia_romana/miscellanea/cleopatra/bust.html; J. Tyldesley, *Cleopatra: Last Queen of Egypt* (New York: Basic Books, 2008).

42 O. Taubman-Ben-Ari, "Intimacy and risky sexual behavior: what does it have to do with death," *Death Studies* 28 (2004), 865–87.

43 D. R. Greenberg and D. J. LaPorte, "Racial difference in body type preferences of men for women," *International Journal of Eating Disorders* 19 (1996), 275–78.

44 According to the 2005 revision of the UN World Urbanization Prospects report, 60 percent of the world's population is forecast to live in cities by 2030. United Nations Department of Economic and Social Affairs Population Division, "World Urbanization Prospects: The 2005 Revision," http://www.un.org/esa/population/publications/WUP2005/2005wup.htm.

45 R. Uher et al., "Functional neuroanatomy of body shape perception in healthy and eating disorder women," *Biological Psychiatry* 58 (2005), 990–97.

46 S. Pinker, *The Better Angels of Our Nature: How We Became Less Violent* (New York: Viking, 2011). Personal communication at the Association of Psychological Science meeting, Boston, May 28, 2010.

47 G. Hirschberger, T. Ein-Dor, and S. Almakias, "The self-protective altruist: terror management and the ambivalent nature of prosocial behavior," *Personality and Social Psychology Bulletin* 34 (2008), 666–78.

48 J. L. Goldenberg et al., "I am not an animal: mortality salience, disgust and the denial of human creatureliness," *Journal of Experimental Psychology: General* 130 (2001), 427–35.

49 Ibid.

50 C. R. Cox, J. L. Goldenberg, T. Pyszcynski, and D. Weise, "Disgust, creatureliness and the accessibility of death-related thoughts," *European Journal of Social Psychology* 37 (2007), 494–507.

51 J. Schimel, J. Hayes, T. Williams, and J. Jahrig, "Is death really the worm at the core? Converging evidence that worldview threat increases death thought accessibility," *Journal of Personality and Social Psychology* 92 (2007), 789–803.

52 J. Greenberg et al., "Sympathy for the devil: evidence that reminding whites of their mortality promotes more favorable reactions to white racists," *Motivation and Emotion* 25 (2001), 113, 133; L. J. Renkema, D. A. Stapel, M. Maringer, and N. W. van Yperen, "Terror management and stereotyping: why do people stereotype when mortality is salient?" *Personality and Social Psychology Bulletin* 34 (2008), 553–64.

53 E. Jonas, J. Schimel, J. Greenberg, and T. Pyszcynski, "The scrooge effect: evidence that mortality salience increases prosocial attitudes and behavior," *Personality and Social Psychology Bulletin* 28 (2002), 1342–53.

54 Ibid.

55 C. R. Kaiser, S. B. Vick, and B. Major, "A prospective investigation of the relationship between just-world beliefs and the desire for revenge after September 11, 2001," *Psychological Science* 15 (2004), 503–6.

56 E. A. Henry, B. D. Bartholow, and J. Arndt, "Death on the brain: effects of mortality salience on the neural correlates of ingroup and outgroup categorization," *SCAN* 5 (2010), 77–87.

57 E. Becker, *The Denial of Death* (New York: Simon & Schuster, 1973). Becker's theories were influenced by the writings of philosophers such as Hegel and Schopenhauer and psychanalysts including Sigmund Freud and Otto Rank.

58 See S. Solomon, J. Greeberg, and T. Pyszczynski, "A terror management theory of social behavior," in M. Zanna, ed., *Advances in Experimental Social Psychology*, vol. 24 (San Diego: Academic Press, 1991).

59 J. Haidt, C. McCauley, and P. Rozin, "Individual differences in sensitivity to disgust: a scale sampling seven domains of disgust elicitors," *Personality and Individual Differences* 16 (1994), 701–13.

60 Goldenberg et al., "I am not an animal."

61 Ibid.

CHAPTER 6: HORROR SHOW

1 *The Exorcist*, adjusted gross, $793,883,100. *Jaws*, adjusted gross, $919,605,900. *Gone With the Wind*, adjusted gross, $1,450,680,400. From

http//www.squidoo.com/usa_all_time_box_office. *Gone With the Wind* (1939) cashes in as the most profitable film to date.

2 R. Mian, G. Shelton-Raynor, B. Harkin, and P. Williams, "Observing a fictitious stressful event: haematological changes, including circulating leukocyte activation," *Stress* 6 (2003), 41–47.

3 http://eric.b.olsen.tripod.com/siodmak.html.

4 S. King, *Danse Macabre* (New York: Berkley Books, 1981), 28.

5 *Horror Movies Crank Up The Gore*, NPR, October 29, 2007. Neda Ulaby "Extreme horror: Basic Escapism or Simply base?"

6 C. McCauley, "When screen violence is not attractive," in J. H. Goldstein, ed., *Why We Watch: The Attractions of Violent Entertainment* (New York: Oxford University Press, 1998), 144–62.

7 *Halloween* is also ranked by critics as one of the best horror films ever.

8 http://en.wikipedia.org/wiki/List_of_horror_films:_1980s.

9 http://en.wikipedia.org/wiki/List_of_horror_films:_2010.

10 L. A. Gilmore, and M. A. Campbell, "Scared but loving it: children's enjoyment of fear as a diagnostic marker of anxiety?" *Australian Educational and Developmental Psychologist* 25 (2008), 24–31.

11 L. Bouzereau, *The Alfred Hitchcock Quote Book* (Secaucus, NJ: Citadel Press, 1993).

12 J. M. Turley and A. P. Derdeyn, "Use of a horror film in psychotherapy," *Journal of the American Academy of Child and Adolescent Psychiatry* 29 (1990), 942–45.

13 B. Ballon and M. Leszcz, "Horror films: tales to master terror or shapers of trauma?" *American Journal of Psychotherapy* 61 (2007), 211–30.

14 D. D. Johnston, "Adolescents' motivation for viewing graphic horror," *Human Communication Research* 21 (1995), 522–52.

15 D. Zillmann, J. B. Weaver, N. Mundorf, and C. F. Aust, "Effects of an opposite-gender companion's affect to horror on distress, delight, and attraction," *Journal of Personality and Social Psychology* 51 (1986), 586–94.

16 Although the adherence to and attractiveness of many gender stereotypes has decreased in recent years, "masculine" and "feminine" traits are still prized for males and females, respectively (see chapter 7, page 180).

17 F. Molitor and B. S. Sapolsky "Sex, violence, and victimization in slasher films," *Journal of Broadcasting and Electronic Media* 37 (1993), 233–42.

18 D. L. Mosher, "Sex differences, sex experience, sex guilt and explicit sexual films," *Journal of Social Issues* 29 (1973), 95–112.

19 M. Oliver, "Contributions of sexual portrayals to viewer's responses to graphic horror," *Journal of Broadcasting and Electronic Media* 38 (1994), 1–17.

20 Sexual arousal was measured by vaginal blood flow and lubrication. D. S. Fleischman, L. D. Hamilton, D. M. Fessler, and C. Meston, "The interactions of disgust and sexual arousing in women: competition between mating and disease avoidance motivations," Human Behavior and Evolution Society, 22d Annual Meeting, June 2010.

21 T. Straube et al., "Neural representation of anxiety and personality during exposure to anxiety-provoking and neutral scenes from scary movies," Human Brain Mapping 31 (2010), 36–47.

22 D. Dutton, The Art Instinct: Beauty, Pleasure and Human Evolution (New York: Bloomsbury, 2009).

23 BBC News: http://news.bbc.co.uk/2/hi/7925719.stm.

24 Identities have been changed.

25 K. Foxhall, "Suicide by profession: lots of confusion, inconclusive data," American Psychological Association, Monitor 32(1) (January, 2001), 19.

26 E. Agerbo et al., "Suicide and occupation: the impact of socio-economic, demographic and psychiatric differences," Psychological Medicine 37 (2007), 1131–40.

27 Y. Barak et al., "Increased risk of attempted suicide among aging Holocaust survivors," American Journal of Geriatric Psychiatry 13 (2005), 701–4.

28 Patrice Oppliger is currently a communications professor at Boston University.

29 Beavis and Butthead is an animated American TV show that ran from 1993 to 1997. The show revels in disgust and potty humor, being obnoxious, dim-witted, and bucking convention.

30 P. A. Oppliger and D. Zillmann, "Disgust in humor: Its appeal to adolescents," Humor 10 (1997), 421–37.

31 D. Zillmann, "Transfer of excitation in emotional behavior," in J. T. Cacioppo and R. E. Petty, eds., Socio-Psychophysiology: A Sourcebook (New York: Guilford, 1983).

32 www.brainyquotes.com.

CHAPTER 7: LUST AND DISGUST

1 D. Hoppe, Healthy Sex Drive, Healthy You: What Your Libido Reveals about Your Life (Encinitas, CA: Health Reflections Press, 2010).

2 J. W. Critelli and J. M. Bivona, "Women's erotic rape fantasies: an evaluation of theory and research," Journal of Sex Research 45 (2008), 57–70.

3 R. Stevenson, T. Case, and M. Oaten, "Effect of self-reported sexual arousal on responses to sex-related and non-sex related disgust cues," Archives of Sexual Behavior 40 (2011), 79–85.

4 S. Karama et al., "Areas of brain activation in males and females during viewing of erotic film excerpts," *Human Brain Mapping* 16 (2002), 1–13.

5 Sadomasochism is a form of sexual role playing where one person gets erotic pleasure from inflicting pain and humiliation and the other gets erotic pleasure from receiving pain and humiliation.

6 R. Stark et al., "Erotic and disgust-inducing pictures: differences in the hemodynamic responses of the brain," *Biological Psychology* 70 (2005), 19–29.

7 Fleischman, Hamilton, Fessler, and Meston, "The interactions of disgust and sexual arousing in women."

8 The 5.5 billion dollar amount was not verified beyond several internet sources.

9 See *The Dream of the Fisherman's Wife* (1814) by Katsushika Hokusai.

10 Bobby Finstock on 07/20/06; http://pointlessbanter.net/2006/07/20/i-bet-you-cant-watch-all-four-minutes/.

11 "Japan Pledges To Halt Production Of Weirdo Porn That Makes People Puke," *The Onion* 45–07 (February 10, 2009)

12 M. Milburn, R. Mather, and S. Conrad, "The effects of viewing R-rated movie scenes that objectify women on perceptions of date rape," *Sex Roles* 43 (2000), 645–64; G. Hald, N. Malamuth, and C. Yuen, "Pornography and attitudes supporting violence against women: revisiting the relationship in nonexperimental studies," *Aggressive Behavior* 36 (2010), 14–20.

13 P. Rozin, and D. Schiller, "The nature and acquisition of a preference for chili pepper by humans," *Motivation and Emotion* 4 (1980), 77–101; Rozin, Haidt, and McCauley, "Disgust."

14 Becker, *The Denial of Death*, 163.

15 J. L. Goldenberg et al., "Understanding human ambivalence about sex: the effects of stripping sex of meaning," *Journal of Sex Research* 39 (2002), 310–20.

16 Ibid.

17 J. L. Goldenberg et al., "Death, sex, love and neuroticism: why is sex such a problem," *Journal of Personality and Social Psychology* 77 (1999), 1173–87.

18 B. Kahr, "The history of sexuality: from polymorphous perversity to modern genital love," *Journal of Psychohistory* 26 (1999), 764–78.

19 Goldenberg et al., "I am not an animal."

20 L. M. Liao and S. M. Creighton, "Requests for cosmetic genitoplasty: how should healthcare providers respond?" *British Medical Journal* 334 (2007), 1090–92.

21 S. W. Gangestad and D. M. Buss, "Pathogen prevalence and human mate preferences," *Ethology and Sociobiology* 14 (1993), 89–96.

22 L. M. DeBruine et al., "Women's preferences for masculinity in male faces are predicted by pathogen disgust, but not by moral or sexual disgust," *Evolution and Human Behavior* 31 (2010), 69–74.

23 A. C. Little, L. M. DeBruine, and B. C. Jones, "Exposure to visual cues of pathogen contagion changes preferences for masculinity and symmetry in opposite sex faces," *Proceeding of the Royal Society B* 278 (2011), 2032–39.

24 J. M. Tybur, V. Griskevicius, and D. Lieberman, "Microbes, mating and morality: Individual differences in three functional domains of disgust," *Journal of Personality and Social Psychology* 97 (2009), 103–22.

25 MHC is a term borrowed from the field of animal immune system genetics, though it is colloquially used to refer to human immune system genetics as well. Technically, human immune system genetics are discussed in terms of human leukocyte antigens or HLA

26 M. L. Thomas et al., "HLA sharing and spontaneous abortion in humans," *American Journal of Obstetrics and Gynecology* 151 (1985), 1053–58.

27 For an in-depth discussion of odor and sex see Herz, *The Scent of Desire.*

28 R. S. Herz and M. Inzlicht, "Gender differences in response to physical and social signals involved in human mate selection: the importance of smell for women," *Evolution and Human Behavior* 23 (2002), 359–64. R. S. Herz and E. D. Cahill, "Differential use of sensory information in sexual behavior as a function of gender," *Human Nature* 8 (1997), 275–86.

29 D. Penn and W. Potts, "MHC-disassortative mating preferences reversed by cross-fostering," *Proceedings of the Royal Society B* 265 (1998), 1299–1306.

30 Kibbutzim are communal agrarian settlements in Israel where children live separately from parents. Data on 2,769 kibbutz marriages found that none were between people who had been reared together during the first six years of life. J. Shepher, *Incest: A Biosocial View* (New York: Academic Press, 1983).

31 D. Lieberman, J. Tooby, and L. Cosmides, "Does morality have a biological basis? An empirical test of the factors governing moral sentiments relating to incest," *Proceedings of the Royal Society B* 270 (2003), 819–26.

32 Sukbir Siwach, "India: Incest is a way of life in Hindu Haryana," *The Times of India,* August 28, 2010; Moin Ansari, *Pakistan Daily,* August 29, 2010.

33 S. Schiff, *Cleopatra: A Life* (New York: Little, Brown, 2010).

34 Leviticus 15:19, 20, 24.

35 L. Bennett, *Dangerous Wives and Sacred Sisters* (New York: Columbia University Press, 1983).

36 J. Delaney, J. Lupton, and E. Toth, *The Curse: A Cultural History of Menstruation* (Urbana: University of Illinois Press, 1988).

37 "The Tampax Report," Ruder, Finn & Rotman, New York, 1981.

38 P. Rozin et al., "Individual differences in disgust sensitivity: comparison and evaluations of a paper-and-pencil versus behavioral measures," *Journal of Research in Personality* 33 (1999), 330–51.

39 T-A. Roberts, J. L. Goldenberg, C. Power, and T. Pyszczynski, "Feminine protection: the effect of menstruation on attitudes towards women," *Psychology of Women Quarterly* 26 (2002), 131–39.

40 L. Meerabeau, "The management of embarrassment and sexuality in health care," *Journal of Advanced Nursing* 29 (1999), 1507–13.

41 J. K. Rempel and B. Baumgartner, "The relationship between attitudes toward menstruation and sexual attitudes, desire, and behavior in women," *Archives of Sexual Behavior* 32 (2003), 155–63.

42 E. Jones, *On the Nightmare* (New York: Liveright, 1931), 109–12.

43 From Saint Alfonso Maria de' Liguori, *Theologia Moralis III*, discussed in A. A. Brill, "Necrophilia," *Journal of Criminal Psychopathology* 3 (1941), 433–53.

44 http://handleonthelaw.com/law-articles/criminal-law-articles/necrophilia-laws-you-are-dying-to-know-about.html.

45 J. P. Rosman and P. J. Resnick, "Sexual attraction to corpses: a psychiatric review of necrophilia," *Bulletin of the American Academy of Psychiatry and the Law* 17 (1989), 153–63.

46 F. S. Klaf and W. Brown, "Necrophilia, brief review and case report," *Psychiatric Quarterly* 32 (1958), 645–52; Rosman and Resnick, "Sexual attraction to corpses."

47 Edgar Allan Poe, "Annabel Lee," first published in *Sartain's Union Magazine*, 1849.

48 L. E. Bonaparte, "Deuil, necrophile et sadisme," *Revue Française de Psychoanalysis* 4 (1930–31), 716–34.

49 S. Freud and J. Strachey, *Three Essays on the Theory of Sexuality* (New York: Basic Books, 1962).

50 William Faulkner, "A Rose for Emily," *Forum*, April 30, 1930. *Family Guy*, "Death is a Bitch," March 21, 2000.

51 *State of Wisconsin v. Grunke*, Wisconsin Supreme Court, 2008.

52 A. L. Mahaffey, A. Bryan, and K. E. Hutchison, "Sex differences in affective responses to homoerotic stimuli: evidence for an unconscious bias among heterosexual men, but not heterosexual women," *Archives of Sexual Behavior* 34 (2005), 537–45.

53 S. L. Bem, "The measurement of psychological androgyny," *Journal of Consulting and Clinical Psychology* 42 (1974), 155–62.

54 C. L. Holt and J. B. Ellis, "Assessing the current validity of the Bem Sex-Role Inventory," *Sex Roles* 39 (1998), 929–41.

55 A. Chandra, W. D. Mosher, C. Copen, and C. Sionean, "Sexual behavior, sexual attraction, and sexual identity in the United States: data from the 2006–2008 national survey of family growth," *National Health Statistics Reports*, Number 36 (Hyattsville, MD: National Center for Health Statistics, 2011).

56 M. E. Kite, "Do heterosexual men and women differ in their attitudes toward homosexuality? A conceptual and methodological analysis," in L. D. Garnets and D. Kimmel, eds., *Psychological Perspectives on Lesbian, Gay and Bisexual Experiences*, 2d ed (New York: Columbia University Press, 2003), 165–87.

57 Sodomy is a catchall legal term for sex acts that are "unnatural." It typically refers to anal sex but may encompass oral sex and bestiality.

58 M. C. Nussbaum, *Hiding From Humanity: Disgust, Shame and the Law* (Princeton, NJ: Princeton University Press, 2004).

59 R. Ellmann, *Oscar Wilde* (New York: Vintage, 1988).

60 Death by hanging was the punishment for sodomy in England until 1861.

61 http://www.websters-online-dictionary.org/definitions/pederasty.

62 "The Dancing Boys of Afghanistan," *PBS Frontline News*, April 20, 2010, http://www.pbs.org/wgbh/pages/frontline/dancingboys/; http://www.foxnews.com/politics/2010/01/28/afghan-men-struggle-sexual-identity-study-finds/.

63 B. O. Olatunji, "Disgust, scrupulosity and conservative attitudes about sex: evidence of a meditational model of homophobia," *Journal of Research in Personality* 42 (2008), 1364–69.

64 M. Daly and M. Wilson, *Sex, Evolution, and Behavior*, 2d ed (Boston: Willard Grant Press, 1983).

65 B. E. Whitley, "Gender-role variables and attitudes toward homosexuality," *Sex Roles* 45 (2002), 691–721.

CHAPTER 8: LAW AND ORDER

1 H. Chua-Eoan, "That's Not a Scarecrow," *Time*, October 19, 1998.

2 http://en.wikipedia.org/wiki/Matthew_Shepard.

3 R. B. Mison, "Homophobia in manslaughter: The homosexual advance as insufficient provocation," *California Law Review* 80 (1992), 133–78.

4 If the case had gone to trial, the Masters and Johnson Institute would have testified that Reubens's handedness (Reubens is right-handed) made

it nearly impossible that he could have been masturbating as alleged, but Reubens didn't want all the added ridicule of a trial and so pleaded "no contest." He served a few weeks of community service. See full interview in *Playboy*, September 2010.

5 Y. Inbar, D. A. Pizarro, J. Knobe, and P. Bloom, "Disgust sensitivity predicts intuitive disapproval of gays," *Emotion* 9 (2009), 435–39.

6 G. Hodson and K. Costello, "Interpersonal disgust, ideological orientations and dehumanization as predictors of intergroup attitudes," *Psychological Science* 18 (2007), 691–98.

7 Olantunji, "Disgust, scrupulosity and conservative attitudes about sex."

8 Y. Inbar, D. A. Pizarro, and P. Bloom, "Conservatives are more easily disgusted than liberals," *Cognition and Emotion* 23 (2009), 714–25.

9 Haidt and Graham, "When morality opposes justice."

10 Inbar, Pizarro, and Bloom, "Conservatives are more easily disgusted than liberals."

11 Information for the Meiwes case came from several sources: http://www.guardian.co.uk/world/2004/jan/31/germany.lukeharding/print, http://newsvote.bbc.co.uk/mpapps/pagetools/print/news.bbc.co.uk/2/hi/europe/3286721.stm, and http://en.wikipedia.org/wiki/Armin_Meiwes.

12 T. Wheatley and J. Haidt, "Hypnotic disgust makes moral judgments more severe," *Psychological Science* 16 (2005), 780–84. The congressman example given has been modified from the original text.

13 S. Schnall, J. Haidt, G. L. Clore, and A. H. Jordan, "Disgust as embodied moral judgment," *Personality and Social Psychology Bulletin* 34 (2008), 1096–1109.

14 J. S. Lerner, D. A. Small, and G. Lowenstein, "Heart strings and purse strings: carryover effects of emotions on economic decisions," *Psychological Science* 15 (2004), 337–41.

15 P. Rozin and L. Singh, "The moralization of cigarette smoking in the United States," *Journal of Consumer Behavior* 8 (1999), 321–37.

16 C.-B. Zhong, and K. Liljenquist, "Washing away your sins," *Science* 313 (2006), 1451–52.

17 S. Schnall, J. Benton, and S. Harvey, "With a clean conscience: cleanliness reduces the severity of moral judgments," *Psychological Science* 19 (2008), 1219-22.

18 K. Liljenquist, C.-B. Zhong, and A. D. Galinsky, "The smell of virtue: clean scents promote reciprocity and charity," *Psychological Science* 21 (2010), 381–83.

19 C.-B. Zhong and K. Liljenquist, "Washing away your sins," *Science* 313 (2006), 1451–52.

20 E. G. Helzer and D. A. Pizarro, "Dirty Liberals! Reminders of physical

cleanliness influence moral and political attitudes," *Psychological Science* 22 (2011), 517–22.

21 S. W. S. Lee and N. Schwarz, "Dirty hands and dirty mouths," *Psychological Science* 21 (2010), 1423–25.

22 Schnall, Haidt, Clore, and Jordan, "Disgust as embodied moral judgment."

23 J. D. Laird and M. Crosby, "Individual differences in self-attribution of emotion," in H. London and R. Nisbett, eds., *Thinking and Feeling: The Cognitive Alteration of Feeling States* (Chicago: Aldine-Atherton, 1974).

24 L. E. Williams and J. A. Bargh, "Experiencing physical warmth promotes interpersonal warmth," *Science* 322 (2008), 606–7.

25 X. Li, L. Wei, and D. Soman, "Sealing the emotions genie: the effects of physical enclosure on psychological closure," *Psychological Science* 21 (2010), 1047–50.

26 Extrapolated from R. Gutierrez and R. Giner-Sorolla, "Anger, disgust and presumption of harm as reaction to taboo-breaking behavior," *Emotion* 7 (2007), 853–68.

27 Ibid.

28 J. Haidt, S. H. Killer, and M. G. Dias, "Affect, culture and morality, or is it wrong to eat your dog?" *Journal of Personality and Social Psychology* 65 (1993), 613–28.

29 L. R. Kass, "The wisdom of repugnance," *New Republic* 216 issue #22, June 2, 1997, 17–26.

30 Ibid.

31 M. Martin, "Anatomist sells body parts online," Reuters, October 22, 2010, http://www.reuters.com/article/idUSTRE69L3G420101022?utm_source=feedburner&utm_medium=feed&utm_campaign=Feed%3A+reuters%2FoddlyEnoughNews+%28News+%2F+US+%2F+Oddly+Enough%29.

32 E. J. Horberg, C. Oveis, D. Keltner, and A. B. Cohen, "Disgust and the moralization of purity," *Journal of Personality and Social Psychology* 97 (2009), 963–76.

33 P. Rozin, L. Lowery, S. Imada, and J. Haidt, "The CAD triad hypothesis," *Journal of Personality and Social Psychology* 76 (1999), 574–86.

34 R. L. Nabi, "The theoretical versus the lay meaning of disgust: implication for emotion research," *Cognition and Emotion* 16 (2002), 695–703.

35 R. S. Herz and A. Hinds, "Stealing isn't gross: language demonstrates how moral violations do not elicit visceral disgust," in preparation.

36 P. Bloom, *Descartes' Baby: How the Science of Child Development Explains What Makes Us Human* (New York: Basic Books, 2004).

37 J. S. Borg, D. Lieberman, and K. A. Kiehl, "Infection, incest, and

iniquity: investigating the neural correlates of disgust and morality,"
Journal of Cognitive Neuroscience 20 (2008), 1529–46.

38 H. A. Chapman, D. A. Kim, J. M. Susskind, and A. K. Anderson,
"In bad taste: evidence for the oral origins of moral disgust," *Science* 32
(2009), 1222–26.

39 Nabi, "The theoretical versus the lay meaning of disgust."

40 R. S. Herz, "PROP taste sensitivity is related to visceral but not moral
disgust," *Chemosensory Perception* 4 (2011), 72–79.

41 The PROP test is the standard test used to assess bitter taste sensitivity and was developed by Linda Bartoshuk at Yale University. See
L. M. Bartoshuk, V. B. Duffy, and I. J. Miller, "PTC/PROP tasting:
anatomy, psychophysics, and sex effects," *Physiology and Behavior* 56
(1994), 1165–71.

42 G. Bell and H.-J. Song, "Genetic basis for 6-n-propylthiouracil taster
and supertaster status determined across cultures," in J. Prescott and
B. J. Tepper, eds., *Genetic Variation in Taste Sensitivity* (New York: Mar-
cel Dekker, 2004).

43 J. E. Mangold et al., "Bitter taste receptor gene polymorphisms are an
important factor in the development of nicotine dependence in African
Americans," *Journal of Medical Genetics* 45 (2008), 578–82.

44 J. M. Tybur, V. Griskevicius, and D. Lieberman, "Microbes, mating and
morality: Individual differences in three functional domains of disgust,"
Journal of Personality and Social Psychology 97 (2009), 103–22.

CHAPTER 9: DISGUST LESSONS

1 P. Lieberman and D. Pizarro, "All Politics is Olfactory," *New York Times*,
October 23, 2010.

2 Liljenquist, Zhong, and Galinsky, "The smell of virtue."

3 World Health Organization, Smoking Statistics, May 28, 2002, http://
www.wpro.who.int/media_centre/fact_sheets/fs_20020528.htm;
P. M. Lantz, "Smoking on the rise among young adults: implications for
research and policy," *Tobacco Control* 12 (Suppl 1) (2003), i60–70.

4 Mathieu-Robert Sauvé, "Le cerveau déteste les images antitabac sur les
paquets de cigarettes," Université de Montréal, *Forum* 40(17), 23 janvier
2006.

5 Warning label displayed on Canadian cigarette packaging to accompany
graph of deaths by various causes.

6 See Rozin and Singh, "The moralization of cigarette smoking in the
United States."

7 There is no scientific evidence to support this as yet, but it may be that gruesome imagery helps deter new would-be smokers who are not yet addicted.

8 D. A. Small and N. M. Verrochi, "The face of need: facial emotion expression on charity advertisements," *Journal of Marketing Research* 46 (2009), 777–87.

9 Todd Hendricks, Senior Vice President—Development ASPCA, personal communication, May 16, 2011.

10 Liljenquist, Zhong, and Galinsky, "The smell of virtue."

11 G. Hirschberger, T. Ein-Dor, and S. Almakias, "The self-protective altruist: terror management and the ambivalent nature of prosocial behavior, *Personality and Social Psychology Bulletin* 34 (2008), 666–78.

12 P. Rozin, "Hedonic 'adaptation': specific habituation to disgust/death elicitors as a result of dissecting a cadaver," *Judgment and Decision Making* 3 (2008), 191–94.

13 See Herz, *The Scent of Desire.*

14 P. Rozin et al., "The borders of the self: contamination sensitivity and potency of the body apertures and other body parts," *Journal of Research in Personality* 29 (1995), 318–40.

15 W. Olesker and L. Balter, "Sex and empathy," *Journal of Counseling Psychology* 19 (1972), 559–62.

16 M. Jabbi, M. Swart, and C. Keysers, "Empathy for positive and negative emotions in the gustatory cortex," *Neuroimage* 34 (2007), 1744–53.

17 E. Valentini, "The role of the anterior insula and anterior cingulate in empathy and pain," *Neurophysiology* 104 (2010), 584–85.

18 T. Singer and C. Lamm, "The social neuroscience of empathy," *Annals of the New York Academy of Science* 1156 (2009), 81–96; J. Decety, "Dissecting the neural mechanisms mediating empathy," *Emotion Review* 3 (2011), 92–108.

19 H. D. Critchley et al., "Neural systems supporting interoceptive awareness," *Nature Neuroscience* 7 (2004), 189–95.

20 J. S. Snowdon et al., "Social cognition in frontotemporal dementia and Huntington's disease," *Neuropsychologia* 41 (2003), 688–701.

21 E. A. Shirtcliff et al., "Neurobiology of empathy and callousness: implication for the development of antisocial behavior," *Behavioral Sciences and the Law* 27 (2009), 137–71.

22 P. Sterzer, C. Stadler, F. Poustka, and A. Kleinschmidt, "A structural neural deficit in adolescents with conduct disorder and its association with lack of empathy," *Neuroimage* 37 (2007), 335–42.

23 D. D. Johnston, "Adolescents' motivation for viewing graphic horror," *Human Communication Research* 21 (1995), 522–52.

24 R. T. Salekin, R. Rogers, K. L. Ustad, and K. W. Sewell, "Psychopathy and recidivism among female inmates," *Law and Human Behavior* 22 (1998), 109–28.

25 K. P. Cosgrove, C. M. Mazure, and J. K. Staley, "Evolving knowledge of sex differences in brain structure, function and chemistry," *Biological Psychiatry* 62 (2007), 847–55.

26 T. Singer and C. Lamm, "The social neuroscience of empathy," *Annals of the New York Academy of Sciences* 1156 (2009), 81–96.

27 See N. R. Craik, J. F. Casas, and M. Mosher, "Relational and overt aggression in pre-school," *Developmental Psychology* 33 (1997) 579–88; N. H. Hess and E. H. Hagen, "Sex differences in indirect aggression: psychological evidence from young adults," *Evolution and Human Behavior* 27 (2006), 231–45.

28 F. B. M. De Waal and A. van Roosmalen, "Reconciliation and consolation among chimpanzees," *Behavioral Ecology and Sociobiology* 5 (1979), 55–66.

29 J. Decety, "Dissecting the neural mechanisms mediating empathy," *Emotion Review* 3 (2011), 92–108.

30 Y. Cheng et al., "Expertise modulates the perception of pain in others," *Current Biology* 17 (2007), 1708–13.

31 Interview with Dr. Jonathan Mirmen, Centers for Disease Control and Prevention; *American Edge*, CNN, June 1997. Transcript available at http://www.anapsid.org/mirmen.html.

32 Cambra, "Snakes on a plane."

33 L. T. Harris and S. T. Fiske, "Dehumanizing the lowest of the low: neuroimaging responses to extreme out-groups," *Psychological Science* 17 (2006), 847–53.

Illustration Credits

Figure 1.1: Woodlouse photographs by Alvesgaspar, available at http://commons.wikimedia.org/wiki/File:Woodlouse_poster.jpg. Shrimp photograph by Rachel Herz.

Figure 1.2: Photograph by Nate "Igor" Smith, available at http://commons.wikimedia.org/wiki/File:Sonya-thomas.jpg

Figure 2.1: Photograph by Rachel Herz.

Figure 2.2: Hieronymus Bosch, *The Garden of Earthly Delights*, c. 1500, detail.

Figure 3.1: Original lithograph plate from *Gray's Anatomy*, 1918.

Figure 4.1: Photograph from the Centers for Disease Control and Prevention's Public Health Image Library.

Figure 5.1: Photograph owned by Rachel Herz.

Figure 5.2: Archives of Institute of National Remembrance, Poland.

Figure 9.1: Photograph by Rachel Herz.

Index

Page numbers in *italics* refer to illustrations.
Page numbers beginning with 237 refer to notes.